COVERED BRIDGES
of
ALABAMA

COVERED BRIDGES

of

ALABAMA

WIL ELRICK & KELLY KAZEK

THE
History
PRESS

"The covered bridge is one of the last surviving icons of rural America."
—*Brian J McKee,* Historic American Covered Bridges

"Covered bridges evoke romantic images of an era when life was slower, smaller and more intimate."
—*Kim Grant,* Backroads of New England

"Few items in America's landscape evoke the nation's past more than a covered bridge."
—National Geographic *178 (1990)*

Published by The History Press
Charleston, SC
www.historypress.com

Cover images: Top front, Pumpkin Hollow Covered Bridge by Josh Box; bottom front, Saunders Family Covered Bridge by Chris McCune and Ben Richardson; top back, Cambron Covered Bridge by Wil Elrick; and bottom back, Poole Covered Bridge by Grady Allen.

First published 2018

Manufactured in the United States

ISBN 9781467140768

Library of Congress Control Number: 2018945785

This book is dedicated to those who help maintain Alabama's remaining covered bridges, and to those who are building them for future generations.

CONTENTS

CONTENTS

PREFACE

For more than two centuries, Americans have had a love affair with covered bridges, those romantic spans once pivotal to transportation and the source of numerous legends. No other type of architecture is more illustrative of life in nineteenth-century America. The bridges were also a source of pride, examples of superior American engineering. Outside the United States, covered bridges only numbered in the hundreds, while in this country, as many as fourteen thousand were built at the height of the bridge-building era.

Covered bridges continue to fascinate us. Societies are dedicated to their preservation, festivals are held in their honor, books are written about them and they are the subjects of movie plots. They have become national treasures. Each year, millions of people will visit one of the estimated 880 remaining covered bridges in the United States.

In today's fast-paced, ever-connected society, it is easy to see the lure of these classic structures that harken to a more leisurely lifestyle. In their early days, covered bridges were places for courting, community gatherings and even weddings, a tradition that has seen a resurgence with modern-day brides and grooms.

Arguably, the most well-known covered bridge in the United States today is the Roseman Covered Bridge in Winterset, Iowa. This bridge, built in 1883, is one of the main characters in Robert James Waller's best-selling novel *The Bridges of Madison County*. When the book was made into a movie starring Clint Eastwood and Meryl Streep, some scenes were filmed on location at

Lidy Walker, or Lidy's, Covered Bridge, shown in the 1960s or 1970s, was a 50-foot span built in 1926 over Big Branch in Blount County. It collapsed in August 2001. *Courtesy of Alabama Department of Archives and History.*

Roseman Bridge. But it is not the only covered bridge to be featured in American fiction. The bridges have made appearances in film versions of "The Legend of Sleepy Hollow," in Edgar Allan Poe's tale "Never Bet the Devil Your Head" and in the cult classic *Beetlejuice*, to name a few.

In Alabama, a state filled with beautiful historical homes and structures, picturesque covered bridges are plentiful. You may have read that only eleven covered bridges survive in Alabama, but that's not the case. While those are the only authentic or historical bridges, Alabama has a total of more than fifty covered bridges in a variety of sizes and styles, including a sparse, modern bridge designed by architecture students at Auburn University. The most recent was built in 2011, the oldest circa 1850.

Of course, that number doesn't compare with some states. Pennsylvania has the most covered bridges with 213, while Ohio is second with 148. But consider the fact that many states have only two or three covered bridges and you'll realize how lucky Alabamians are to have so many picturesque spans in our state.

The eleven surviving historical bridges are the subject of a documentary by filmmaker and Winfield, Alabama native Max Shores. His film *Bridges to the Past* was shown on Alabama Public Television and "follows a timeline of

Clarkson Covered Bridge, also called Clarkson-Legg Covered Bridge, is a 250-foot-long span built in 1904 in Cullman County. *Photo by Wil Elrick.*

Alabama history as witnessed by bridges which have withstood the tests of time, high water, and heavy loads," Shores said. "Wooden covered bridges played an important role in the development of Alabama by providing safe passage over creeks and rivers at locations where crossing in a horse-drawn wagon or buggy would otherwise be extremely difficult, or impossible. During the nineteenth and early twentieth centuries, hundreds of covered bridges were constructed in the state, but now only a few remain. Those persistent engineering marvels that still stand have become popular tourist attractions, giving visitors insight into the needs and hardships of our ancestors."

Of the more contemporary bridges located in Alabama, a handful are little more than footbridges, decorations on someone's private property, while one bridge built in 2000 is the longest in the state at 334 feet long.

A resurgence of interest in the early 2000s led to the restoration of several historical bridges, while others have been lost to storms or fire. Nectar Covered Bridge, which at the time was the longest in the state at 385 feet, was lost to arson in 1993. Currently, only one of Alabama's historical bridges is endangered: Waldo Covered Bridge. It no longer has approaches and is in danger of collapse, but no agency is charged with maintaining it.

Here are the eleven surviving historical bridges:

- Alamuchee, or Alamuchee-Bellamy, Covered Bridge in Livingston in Sumter County, built in 1861.
- Clarkson Covered Bridge in Cullman County, built in 1904.

- Coldwater Covered Bridge in Oxford in Calhoun County, built circa 1850.
- Easley Covered Bridge in Rosa in Blount County, built in 1927.
- Gilliland Covered Bridge in Gadsden in Etowah County, built in 1889.
- Horton Mill Covered Bridge in Oneonta in Blount County, built in 1934.
- Kymulga Covered Bridge in Childersburg in Talladega County, built in 1861.
- Old Union Crossing Covered Bridge in Mentone in DeKalb County. A portion is thought to have been built circa 1863; it is no longer categorized as "authentic."
- Salem-Shotwell Covered Bridge in Opelika in Lee County, built in 1900.
- Swann Covered Bridge in Cleveland in Blount County, built in 1933.
- Waldo Covered Bridge in Talladega County, built in 1858.

Through the following pages, you will join us on a journey through the past, present and future of Alabama's covered bridges.

Because some of these bridges are small structures on private property, and because bridges are still being built, it would be impossible to catalogue every covered bridge in Alabama. However, we tried to identify as many as possible and give as much data as possible on each. To get this information, we crossed the majority of the state's covered bridges, either on foot or in our cars. We visited rural communities, shimmied down a bluff to find the ruins of an iconic lost bridge and talked with people who built their own bridges and those who are dedicated to their preservation.

The book is organized into sections so that they stand on their own, which means you will come across some repetition of information. We wanted to provide a comprehensive look at these nostalgic structures. We hope you enjoy it.

ACKNOWLEDGEMENTS

When writing a book, we always come across plenty of folks willing to help us share Alabama's remarkable history. In this case, we'd like to thank Don and June Osborne, Claude and Virginia Price, Joe Watts and Paul Franklin with the Alabama Birding Trail, Billy Milstead with RuralSWAlabama.org, the Blount County Memorial Museum, the Alabama Department of Archives and History, Alabama Mosaic and the Encyclopedia of Alabama.

Those who helped by taking photos of bridges are John Trent with Wehle Land Conservation Center, Karrie Jacobs, Glenn Wills, Charlotte Kirkpatrick, Shannon Kazek, Chris McCune, Ben Richardson, Pamela Watts, Grady Allen, Josh Box, Phillip M. Burrow, Charles Byrom and Kathy Shirley.

WHY ARE BRIDGES COVERED?

In any book about covered bridges, one of the first questions to answer is, "Why are covered bridges covered?" The answer is not quite as straightforward as one would think.

A common reason given for covering bridges is to protect the decking or flooring. This, however, is not the most important reason for covering bridges. Firstly, decking is relatively inexpensive, and secondly, decking requires maintenance and replacing after a significant amount of traffic usage, even when it has been protected from the elements by the roof.

In the northern or midwestern states, it is a commonly held belief that bridges were covered to keep snow from piling up on them. Instead of accumulating on the flat surface of the decking and adding more and more weight to the bridge, the snow would simply slide off the slanted roof, keeping the flat surface free and clear of snow. But in the early days of covered bridges, when roadways could not be quickly plowed or cleared of snow, people traded their wheeled carriages for tracked sleds and sleighs so they could cut easily through the snow. One job of bridge tenders of the time was to shovel snow *onto* the decks of covered bridges so that the sleighs would be able to travel on them. This is an instance where the bridge covering would create a problem rather than prevent one.

Many authors of books and articles on covered bridges claim that they were covered to protect travelers and provide them with shelter in case of a storm. In the 1800s and even into the early 1900s, foot traffic was much

Left: Information with this 1939 photo notes, "Wooden approach leading to a covered bridge." No other information was given. *Courtesy of Alabama Department of Archives and History.*

Right: Gilliland's Covered Bridge in Noccalula Falls State Park, shown in the 1970s. *Courtesy of Alabama Department of Archives and History.*

more common than it is today, especially in rural areas, where many covered bridges were built. The theory was that travelers walking on the wet decking of a bridge were more likely to slip and possibly fall to their deaths, and roofs helped to prevent this. While that makes sense, and people likely often took shelter under covered bridges in those days, it is still not the true reason for covering bridges. In reality, bridges didn't provide much protection from severe storms.

Another theory is that covering bridges made crossing the water easier for horses or other livestock being moved by farmers. The covers prevented the animals from seeing the water below and getting spooked. Interestingly, some towns of the day imposed fines on people whose animals crossed bridges too quickly, claiming it added additional wear to the bridge. Still others argue that covering bridges was purely for aesthetics—the covers would hide the support timbers.

In the end, the most important reason for covering bridges has nothing to do with appearances or livestock or snow accumulation. Bridges were covered to protect the main structural parts of the bridge: the trusses. Trusses are expensive, heavy and difficult to put into place. Keeping them safe from the elements was of vital importance because deteriorating trusses endangered the rest of the bridge. Without a covering, the trusses would last ten to fifteen years, but with a covering, they could last for more than a century—and many have. The Federal Highway Administration's *Covered Bridge Manual* notes:

Cambron Covered Bridge was built in 1974 along the Nature Trail atop Huntsville's Green Mountain. *Photo by Wil Elrick.*

Timber bridges initially were built without coverings and failed in just a few years because of rot and deterioration, because chemical wood preservatives were not available or used. Builders familiar with the construction of houses, barns, and large community structures naturally added siding and roofs to help protect the bridge. They understood that the covering would soon pay for itself. They believed that regular maintenance and occasional replacement of the light covering was far easier and cheaper than building an entirely new bridge. North American covered bridges still serve after nearly 200 years, due in part to the continued soundness of the trusses, which was possible only with these protective coverings.

In their heyday, bridges were covered in the United States more than anywhere else in the world. Historians estimate that fourteen thousand bridges were built in the United States between 1820 and 1960, more than quadruple of any other country in the world. A 1936 Associated Press article described Alabama's love affair with covered bridges:

Alabamians began building covered bridges almost as soon as they began building Baptist and Methodist churches and one-room school houses. A

Buzzard Roost Covered Bridge, shown in the 1950s, was a 94-foot span that some claim was built circa 1820, although there are no records to support that date. It was one of the first covered bridges built in Alabama. It burned in 1972. *Courtesy of Alabama Department of Archives and History.*

bridge without a roof was not a thing for a county to be proud of and, besides, when a shower came what were travelers in open buggies to do for shelter? And day and night, wasn't a covered bridge a rendezvous for lovers while youngsters concealed themselves among the rafters to listen in?

ALABAMA'S COVERED BRIDGE TRAIL AND FALL FESTIVALS

A labama has eleven authentic historical bridges and dozens of modern, decorative bridges. Blount County, known as the Covered Bridge Capital of Alabama, is home to three of the state's oldest bridges, including the nation's highest above a waterway. Blount County's bridges have another claim to fame: They are the only three historical bridges that still carry motor traffic. They were closed in 2009 after inspections raised safety concerns and reopened in 2012 after extensive renovations.

Those three bridges and ten others are listed on Alabama's official Covered Bridge Trail. The Alabama Department of Tourism lists the trail in this order:

- Alamuchee-Bellamy Bridge on the University of West Alabama campus in Sumter County was built over the Sucarnoochee River in 1861. It was restored and moved to Alamuchee Creek in 1961.
- Clarkson-Legg Covered Bridge was built in 1904 near Bethel in Cullman County with lattice-style planks. It was restored in 1975.
- Swann Covered Bridge, built circa 1933 north of Cleveland in Blount County, spans 324 feet across the Locust Fork River.
- Easley Covered Bridge was built in 1927 off U.S. Highway 231 south of Rosa in Blount County. The 95-foot-long span has a tin roof.
- Horton Mill Bridge, built in 1935 near Oneonta in Blount County, is the highest covered bridge in the nation, running 70 feet above the Black Warrior River.

A view of Horton Mill Covered Bridge in Oneonta. *Photo by Wil Elrick.*

- Old Union Crossing Bridge at the Shady Grove Dude Ranch near Mentone is 90 feet long and spans the West Fork of Little River. It was moved from Lincoln to its current site in 1972.
- Gilliland Covered Bridge, built in 1899 in Gadsden, was restored in 1968 and moved to Noccalula Falls State Park.
- Coldwater Covered Bridge was constructed in 1850 near Oxford by a former slave. It was moved from Coldwater Creek to its current location at Oxford Lake and Walking Trail in Calhoun County.
- Waldo Covered Bridge was built in the mid-1800s in Talladega County.
- Kymulga Covered Bridge was constructed in 1861 to span Talladega Creek.
- Poole's Covered Bridge was built for the Pioneer Museum of Alabama in Troy.
- Rickard's Mill Park includes a restored gristmill, a syrup mill, a blacksmith shop and a unique covered bridge that also acts as a gift shop.

Gilliland Covered Bridge, sometimes known as Reece Bridge or Gilliland-Reese (*sic*) Bridge, is an 81-foot-long span that was moved from its original location to Gadsden's Noccalula Falls Park in 1967. It was built in 1899 by Jesse Gilliland on his Reece City plantation. This is a view from the back side of the bridge. *Photo by Kelly Kazek.*

- Horace King Memorial Bridge is a replica of the style built by the onetime slave turned architect. Several Horace King bridges once stood in Alabama, but all were eventually demolished. The replica in Valley honors the renowned builder.

COVERED BRIDGE FESTIVALS

Two festivals are held in Blount County each October to celebrate the county's picturesque spans. The Covered Bridge Arts and Music Fest has been held in Oneonta each October since 1983. The free family event features a guided river walk, a car show, a Kids' Zone, live music and food. Festivalgoers can shop at hundreds of arts-and-crafts booths in downtown Oneonta as live music plays. A 5K run and a quilt show are held in conjunction with the event.

Displays at the Blount County Memorial Museum show the history of covered bridges in Blount County. *Photo by Wil Elrick.*

In addition, Oneonta High School's marching band has hosted an annual Covered Bridge Marching Festival the first Saturday in October since 1998. For more information on the Covered Bridge Arts and Music Fest, visit www. facebook.com/CoveredBridgeFest or call the chamber of commerce office at 205-274-2153.

HISTORICAL SIGNIFICANCE

A 1981 application to have the Blount County bridges placed in the National Register of Historic Places explains their importance in the county's history:

> *The bridges are significant in the transportation history of Alabama and Blount County as the last remaining examples of the major nineteenth-century solution to bridging streams and rivers in the state—the covered bridge. Although by the turn of the century, metal truss bridges had replaced the wooden bridge in most parts of the state, Blount County continued to construct covered bridges until 1935. At least 12 covered bridges were constructed in the county during the twentieth century, six of which were constructed in the 1930s. The bridges served to increase trade and communications between the numerous small communities of Blount County.*

At the time of the application, the 1934 Nectar Covered Bridge was still standing. It burned in 1993, a terrible loss because of its historical significance and because it was the longest in the state at 385 feet. The application noted that building materials for the Blount County bridges came from Alabama. The bridges "are all Town Truss bridges constructed of local oak and pine, with stone or concrete piers and abutments. All hardware on the bridges came from Vann and Young Supply Company in Birmingham."

Two members of the Tidwell family played a large role in Blount County's bridges. Easley Covered Bridge was constructed circa 1927 by Forrest Tidwell, a foreman of the country crew. His nephew, Zelma C. Tidwell, learned bridge-building skills from his uncle and went on to construct Swann, Horton Mill, Vaughn, Locust Fork, Tyre Green and Nectar Bridges in Blount County. Only Easley, Swann and Horton Mill are still standing.

The NRHP application noted that Zelma Tidwell chose to use the Town Truss style because "it was the strongest bridge construction." Tidwell was born in 1902 in the community of Locust Fork and was initially a bridge painter for the county. In 1930, he took over his uncle's job as foreman of the

A view of Easley Covered Bridge in Oneonta. *Photo by Wil Elrick.*

Blount County Master Gardeners Memorial Covered Bridge was built in Oneonta's Palisades Park in 2000 to honor local master gardeners. *Photo by Wil Elrick.*

This undated photo shows Nectar Covered Bridge before it was burned in 1993. *Courtesy of Blount County Memorial Museum.*

bridge construction crew. He joined the Alabama Highway Patrol in 1936, after which no other covered bridges were constructed in Blount County.

The Blount County Memorial Museum has displays and photos of the county's three surviving historical covered bridges, along with other local artifacts. The museum at 204 Second Street North in Oneonta is open from 8:00 a.m. to 5:00 p.m. Tuesdays through Thursdays. For information, call 205-625-6905 or visit blountmuseum.org.

HORACE KING

FROM SLAVE TO
RENOWNED BRIDGE BUILDER

Visitors to the Alabama Capitol Building in Montgomery marvel at its most unique feature: two elegantly curving staircases leading to the third floor. The twin cantilevered spiral staircases, built circa 1851, are the work of architect Horace King, a man born into slavery who became a renowned architect best known as a bridge builder.

Eighteen years after the completion of the staircases, King was elected to Alabama's House of Representatives and would have often visited the capitol, using the very steps he designed. Built after King was emancipated by an act of the Alabama legislature in 1846, the jaw-dropping staircases are among the few surviving examples of King's projects that were not bridges.

King built bridges throughout Alabama, Georgia and Mississippi, although none survives in Alabama. A replica, the Horace King Memorial Covered Bridge in Valley, Alabama, celebrates his work.

According to the New Georgia Encyclopedia, "He constructed massive town lattice truss bridges over nearly every major river from the Oconee in Georgia to the Tombigbee in Mississippi and at nearly every crossing of the Chattahoochee River from Carroll County to Fort Gaines."

King was born in 1807 in South Carolina. When he was twenty-three, he was sold to John Godwin, who was a building contractor. King and Godwin formed an unusual bond that would continue after King's emancipation. In 1833, the Godwin family moved to Girard, Alabama, along with King, his mother and siblings. Together, Godwin and King built as many as forty cotton warehouses in Apalachicola, Florida, as well as courthouses in Russell

Horace King was born a slave in South Carolina in 1807. He and his former master, John Godwin, became partners in a construction business and built numerous covered bridges in Alabama, Georgia and Mississippi. None remains in Alabama. *Courtesy of Columbus Museum, Columbus, Georgia.*

and Lee Counties in Alabama, among other non-bridge projects. In 1841, Godwin and King built a covered bridge from Columbus, Georgia, to Girard—thought to be King's first bridge project. He lived with his family for a time in Girard, which is now called Phenix City.

In 1839, King married a free black woman, Frances Gould Thomas, a union that guaranteed freedom for their children.

When Godwin died in 1859, King erected a monument on his grave that is etched with the words, "This stone was placed here by Horace King, in lasting remembrance of the love and gratitude he felt for his lost friend and former master." Godwin is buried at the Godwin Family Cemetery in Phenix City, Alabama.

Despite the fact that King was a Unionist, he was forced to do work for the Confederacy during the Civil War, including construction of a mill to supply wood to the navy, according the Encyclopedia of Alabama. King returned to Columbus in 1863, after which he was conscripted by the Confederacy to build obstructions on the Apalachicola and Alabama Rivers to prevent attacks from the water. Columbus was in an economic boom because of the war; it was a major shipbuilding hub. King was also assigned to help build ships for the Confederate navy at the Columbus Iron Works and Navy Yard. He built a rolling mill, which manufactured the materials to cover ironclad ships. The New Georgia Encyclopedia noted that he also "supplied timbers and erected a major building for the Confederate navy there."

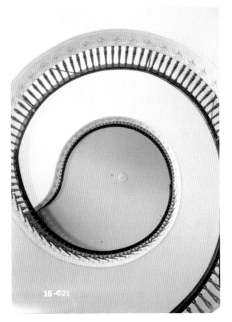

Left: The most eye-catching feature of the Alabama State Capitol Building is the double staircase built by former slave Horace King, shown here in a 1934 photo. The elegantly curving twin cantilevered spiral staircases that lead to the third floor were built circa 1851. *Courtesy of the Historic American Buildings Survey/Library of Congress.*

Right: Another view of the staircase built by Horace King at the Alabama capitol. *Courtesy of the Historic American Buildings Survey/Library of Congress.*

Reportedly, King worked for the Confederacy against his will. Most of the bridges he'd built across the South were burned by Union troops. According to the book *Bridging Deep South Rivers: The Life and Legend of Horace King*, by John S. Lupold and Thomas L. French, he demanded payment for his losses. "After the Civil War, King claimed to be a unionist and petitioned the federal government for damages inflicted on his property by U.S. troops." His request was denied.

The encyclopedia noted, "In recent years he has been cited as a black Confederate, an African American who supported the Southern cause. If his Unionist testimony reflected his true opinion, however, King shunned any association with the Confederacy." During Reconstruction, King rebuilt many structures destroyed in the war, such as railroad bridges, textile mills and cotton warehouses.

Above: People picnicking in Wetumpka in the 1890s, with the bridge built by Horace King in the background. *Courtesy of Alabama Department of Archives and History.*

Left: This bridge over the Chattahoochee River in Eufaula, built by Horace King, is shown being demolished in 1924. *Courtesy of the Alabama Department of Archives and History.*

King's first wife died in 1864, and he married Sarah Jane Jones McManus in 1865. He was elected to serve in the Alabama House of Representatives in 1869 and served two terms, leaving office in 1872 because he wanted to focus on his construction work. The year he left office, King and his family moved to LaGrange, Georgia, where he continued building bridges, stores, houses and college buildings with his sons.

King died on May 28, 1885, and is buried at Stonewall Confederate Cemetery in LaGrange, Georgia. His obituary was published in major newspapers, including those in Atlanta and Columbus. He was survived by three of four sons and a daughter: Washington W. (1843–1910), Marshall Ney (1844–1879), John Thomas (1846–1926), Annie Elizabeth (1848–1919) and George (1850–1899). His sons continued their father's building business. King built an estimated 125 bridges in his lifetime, according to Columbus State University.

DILLingham Street Bridge
Alabama Side
Built 1870
Woodwork - Ex-slave Horace King

Horace King helped build the Dillingham Street Covered Bridge over the Chattahoochee River between Columbus, Georgia and Girard, Alabama (present-day Phenix City), in 1832. *Courtesy of Alabama Department of Archives and History.*

Although some sources claim that King constructed additional covered bridges in Alabama—including in Florence and Tuscaloosa—the following are the ones that are documented as his work:

- Eufaula Bridge, a 540-foot span over the Chattahoochee River. It was built in 1838 and demolished in 1924.
- Dillingham Street Bridge from Columbus, Georgia, to Girard (now Phenix City), Alabama. It was built in 1841. An etching of the Dillingham Bridge appeared on two-dollar notes issued by the Bank of Columbus. The bridge was destroyed in the Battle of Columbus and rebuilt by King. It was replaced in 1912 by a non-covered bridge.
- Wetumpka Bridge spanned the Coosa River. It was built in 1844 and washed away in a flood in 1866.
- Tallassee Bridge was built to span the Tallapoosa River in 1845. Its destruction date is unknown.

Another Alabama bridge, Meadows Mill in Lee County, was built in 1902 by King's son Washington W. King. It was destroyed by arson on October 4, 1973. In honor of Horace King's work on the staircases, his portrait was hung in the Alabama Capitol in February 2017. It is the first portrait of an African American to have a permanent place in the capitol building.

In Georgia, one King-built bridge survives, the Big Red Oak Creek Covered Bridge, also known as Imlac Covered Bridge, near Woodbury in Meriwether County. The bridge, built in 1840, was constructed using 2,500 wooden pegs. It is Georgia's oldest surviving covered bridge. ExploreGeorgia.org notes, "Including approaches, it stretches for 391 feet, making it the longest wooden bridge in Georgia. The main span is 253 feet long and is the state's oldest covered bridge."

Georgia is also home to a bridge built by Washington W. King. Watson Mill Covered Bridge was built in 1888 at a cost of $3,228. The 228-foot-long bridge is located in Watson Mill State Park in Madison County. In Carroll County, Georgia, county officials and the Trust for Public Land are planning to build a replica of a Horace King bridge in his honor. The bridge will be on the site of the 480-foot-long Moore's Mill Covered Bridge, built circa

Meadows Mill Covered Bridge in Lee County was built in 1902 by Horace King's son, W.W. King. It burned in 1973. *Courtesy of the Historic American Buildings Survey/Library of Congress.*

1857 across the Chattahoochee River. It was burned in 1864 and rebuilt by King, but the new span was destroyed by a flood in 1881. It was later replaced with a metal bridge.

According to the Trust for Public Land, "In July 1864, General William Tecumseh Sherman sent Major General George Stoneman to find a crossing over the Chattahoochee. An artillery battle ensued with the Confederate soldiers across the river, and although the Union troops were driven off, they set the bridge afire. This delayed the Battle of Atlanta for several days because Sherman refused to cross the river 40 miles upstream until he knew Stoneman had either secured or destroyed the bridge."

LEGENDS ABOUT
COVERED BRIDGES

Bridges have a mystique that spawns numerous urban legends, perhaps because of the rushing water below and fear of collapse. Tales abound about covered bridges—most of them with common themes not based on fact. Here are a few legends that can be found online or in libraries:

- If a young woman ties a chain of daisies across the entrance to a covered bridge, the first person to break the chain will be her true love.

- The longest covered bridge in the world, the Hartland Bridge in New Brunswick, Canada, is known as the Kissing Bridge. It was built in 1901 and covered in 1920—all 1,282 feet of it. That's when Hartland gained its reputation. A speed limit on the bridge mandated that people couldn't drive their horses at more than a walk. That meant quite a lengthy trip in a dark bridge tunnel, giving suitors time to steal kisses from their girls.

- Several covered bridges are known as "wishing bridges," including Hartland. According to lore, people who make a wish upon entering a covered bridge, with their eyes closed and fingers crossed, will have their wishes granted.

Mountain Oaks Covered Bridge was built in 1970 in Mountain Oaks Estates subdivision in Hoover. *Courtesy of Charlotte Kirkpatrick.*

- Ghost tales are attached to many covered bridges throughout the United States. That's why in some areas the walls of covered bridges are decorated with horseshoes painted red. They supposedly help ward off evil spirits.

- Numerous covered bridges, such as Jericho Covered Bridge in Kingsville, Maryland, are haunted by lynching victims from the Civil War. According to legend, people who stop their cars beneath a bridge at night, with their headlights off and windows rolled down, and then honk three times will see an apparition in their rearview mirrors. The figures of three men will be hanging by ropes from the rafters of the bridge. The drivers' cars won't start again until the apparitions disappear.

- As you'd expect, legends also surround the famed bridge from Washington Irving's story "The Legend of Sleepy Hollow"—despite the fact that it doesn't exist. Anyone who has seen film versions of the story by Walt Disney and Tim Burton will recall the Headless Horseman crossed a covered bridge. In fact, there never was

Top: Pintlala Creek Covered Bridge on Old Selma Road in Montgomery County is shown in 1954. *Courtesy of Alabama Department of Archives and History.*

Middle: Standridge Covered Bridge over the Locust Fork of the Black Warrior River near Hayden in Blount County is shown in 1967. It is no longer standing. *Courtesy of Alabama Department of Archives and History.*

Bottom: Standridge Covered Bridge soon after being destroyed by arson on November 18, 1967. *Courtesy of Alabama Department of Archives and History.*

a covered bridge in the town where the story is set, according to VisitSleepyHollow.com.

- The same site says the Sleepy Hollow bridge depicted in the original story would have been an uncovered wooden span across the Pocantico River that has been gone for decades—perhaps as much as two centuries. "Sleepy Hollow village historian Henry Steiner has documented at least five distinct bridges that carried the Albany Post Road over the stream. Scarcely a trace remains of any…." Local historians place the location of Ichabod's collision in Sleepy Hollow Cemetery, which happens to have a bridge in it (although it is not covered).

- Like many bridges without roofs, covered bridges have been connected to an old urban myth: the Cry Baby Bridge. Lore says cars of motorists crossing certain bridges will feel their cars shake and hear the sounds of a baby crying. The crying supposedly comes from an infant who died in the waters below, from murder, a car accident, a wagon accident from pioneer days or variations on the theme.

- A legend attached to the covered bridge at Camp Nelson, Kentucky, says that it was built during the Civil War in such a manner that it would collapse if one bolt were removed, according to the book *Kentucky's Covered Bridges*. The design would supposedly allow someone to pull the bolt and collapse the bridge so it wouldn't fall into enemy hands. Authors Robert Laughlin and Melissa Jurgensen wrote, "Many a young man enticed his date to give him a forbidden kiss by simply offering that if she did not remove her hesitations, he would remove the bolt."

- Many covered bridges have this legend attached: If a driver puts his car in neutral while on the bridge, invisible hands will push it to the other side.

THE HANGING AT
ALAMUCHEE BRIDGE

Over the years, particularly in the mid-nineteenth century, covered bridges were perceived as a convenient place for lynch mob justice. The beams of the frame made a perfect place to hang a ne'er-do-well who had terrorized the citizenry, plus the bridges were conveniently cloaked in darkness. Tales of hangings are common about covered bridges across the nation, although most fall into the category of urban legend. In Alabama, there is documentation of a lynching at or near one of the state's oldest spans, the Alamuchee Covered Bridge, built in 1861.

The bridge was built on the orders of Confederate army captain William Alexander Campbell Jones, for troops to access Mississippi. Sometimes called Bellamy or Alamuchee-Bellamy, it originally spanned the Sucarnoochee River between Livingston and York.

The man who was reportedly hanged at this bridge was one of the state's most infamous residents, Steve Renfroe, also known as the "Outlaw Sheriff." Renfroe was born in Georgia, circa 1852, and came to the town of Livingston, Alabama, in late 1867 or early 1868. He was a tall, handsome man who rode straight and tall on his white horse, which caught the attention of townspeople. He spoke very little of his past, telling everyone that he was on the run from the federal government for killing some black men in a riot.

Nonetheless, Renfroe was quite personable and made friends quickly in the small town. He eventually fell in with the Ku Klux Klan, where he started a reign of terror against black residents and Union sympathizers. He was indicted in Sumter County for the murder of a magistrate's bodyguard,

Right: A rendering of Steve Renfroe, the "Outlaw Sheriff," who was hanged at the Alamuchee Covered Bridge, according to legend. *From the* Tuscaloosa News*, 1874.*

Below: Alamuchee Covered Bridge was built in 1861. It was moved to the campus of the University of West Alabama in 1971. *Courtesy of Charlotte Kirkpatrick.*

as well as in connection with the disappearance of a local judge. The charges were not pursued, however, and the populace of Livingston and Sumter County voted him sheriff in 1877. Locals reportedly said at the time that there was no more popular man in Livingston than Steve Renfroe, who was the kind of man who basked in the attention.

In his new position as sheriff, Renfroe was initially popular for his hardline stance against carpetbaggers, but then people began to notice changes in the sheriff. He was eventually tied to mismanagement of his office stemming from missing county money and vanished court records. When he was elected for a second term as sheriff, Renfroe's crimes became even bolder. Several counts of robbery, drinking, blackmail, arson and thieving were on the list of offenses being perpetrated by the local sheriff.

Finally, in 1886, the county district attorney had enough of Renfroe's crime spree and secretly presented charges against him to a grand jury, which indicted him on twenty-one charges. Steve Renfroe was immediately arrested. Just as quickly, he broke out of his own jail, stole all the firearms at the Sheriff's Office and released all the prisoners. This was the official beginning of his outlaw life.

Over the next several years, Renfroe was arrested in various towns throughout Alabama, Mississippi and Louisiana, but he always ended up escaping and going into hiding before turning up again to commit more crimes. In July 1886, karma finally caught up with Steve Renfroe in rural Mississippi about sixty miles east of Livingston. A group of locals ambushed him and filled him full of birdshot before taking him to the authorities in Livingston, who were surprised to have the notorious wanted outlaw dropped right into their laps. On July 13, 1886, Renfroe was placed in the Livingston jail once again, and everyone in town knew that he was all too familiar with this jail and would escape again, just as he had previously. He would then unleash more torment on the townsfolk for his latest imprisonment.

Alabama's Outlaw Sheriff would not get the opportunity to escape this time, however; the citizens had had enough. As darkness fell that day, eight chosen men, upstanding citizens who had initially admired and supported the sheriff, walked into the Livingston jail, forcefully took the jailer's keys and removed Renfroe from his cell. Nine years after he had been elected sheriff, Steve Renfroe, the outlaw, was marched through the darkened streets of Livingston with townsfolk quietly watching the silent parade.

That much of the story is well documented. The exact place of Renfroe's hanging remains unclear. One legend says Renfroe was pulled into the Alamuchee Covered Bridge, where the mob threw a rope over its rafters. The other legend says that the noose was strung from the limbs of a chinaberry tree beside the bridge on the water's edge. Either way, Renfroe met his maker at the end of the hangman's rope.

Before the hanging, the posse surrounding the outlaw were reported to have told him, "We are all your friends, Steve. We are doing this for your own good." Some stories of the event say that Renfroe asked his executioners to say a prayer for him because he had never prayed a day in his life.

The sheriff who took Renfroe's place and his deputies arrived at the bridge too late to stop the hanging. They removed the body and returned it to town, but no family members came to claim the remains. No one was ever charged with his hanging. Renfroe's body was buried in an unmarked grave but was later dug up and reinterred at Old Side Cemetery in Sumterville between his first and second wives.

As with any tale of such violence, it left its mark on locals, spawning legends and ghost tales. Some say that the ghost of Steve Renfroe walks back and forth across the bridge each night, desperately trying to get someone to notice him. Another version says that his ghost only returns to the site on the anniversary of his death on July 13. Another legend says that the chinaberry tree beside the bridge will sway when there is no wind, as if the outlaw were still swinging from its branches. For years after the lynching, field hands would reportedly refuse to work near the bridge because they feared the sheriff's ghost.

In 1924, the bridge was moved about five miles from its original site to span Alamuchee Creek on the old Bellamy-Livingston Road. It was used for motor traffic until 1958, when it was abandoned. A historical marker at the site gives the year of construction as 1860. It says, in part: "1860 Captain W.A.C. Jones of Livingston designed and built the bridge of hand-hewn yellow pine put together with large pegs, clear span 88 feet, overhead clearance 14 feet, and inside width 17 feet, across the Sucarnoochee River on old State Road South of Livingston." In 1971, the bridge was moved to the campus of the University of West Alabama, where it spans Duck Pond. Visitors can park in the Student Union parking lot and walk to the bridge.

Although it has not stood near the Sucarnoochee River for nearly one hundred years, legends continue to surround the old bridge, and some still fear crossing it, lest they encounter the ghost of the Outlaw Sheriff.

MILLER BRIDGE WAS ONCE LONGEST IN STATE

Visitors to Horseshoe Bend National Military Park in Tallapoosa County, Alabama, can see the ruins of stone abutments that mark the site of what was once the nation's longest covered bridge. At 600 feet long, Miller's Ferry Bridge, also called Miller Covered Bridge, was a four-span wooden covered bridge built in 1908 across the Tallapoosa River.

The bridge, of Town Lattice Truss design, was constructed by Joe Winn of Dadeville at the site of the ferry once used to cross the river. He charged the county $13,986. More than 1,600 white oak pegs were required in construction of the bridge, which also featured iron-framed approach spans. According to the National Park Service, workers used a temporary derrick and a system of cables to construct the bridge over falsework.

Miller Covered Bridge was a 600-foot span built in 1887 over the Tallapoosa River at New Site. At the time, it was the longest covered bridge in the United States. The bridge was built at the launch point of the old Miller's Ferry, which is now part of Horseshoe Bend National Military Park. The bridge was destroyed during a flood in July 1963. *Courtesy of Alabama Department of Archives and History.*

Miller Covered Bridge was a 600-foot span built in 1887 over the Tallapoosa River at New Site at what is now Horseshoe Bend National Military Park. *Courtesy of Alabama Department of Archives and History.*

In the late 1950s, a concrete bridge was constructed parallel to the covered bridge. Just a few years later, in June 1963, the covered bridge was washed away by floods caused by heavy rains. Horseshoe Bend National Military Park was the site of the last battle fought between Andrew Jackson's troops and the Red Sticks in the Creek Indian War. Visitors to the park can picnic near the bridge ruins.

WAS THIS BRIDGE A CIVIL WAR CROSSING?

O ld Union Crossing Covered Bridge is one of the most unusual in Alabama: it's a hybrid of old and new. But was it built during the Civil War?

The privately owned bridge is located on Lookout Mountain not far from the ski slopes at Cloudmont Ski & Golf Resort off DeKalb County Road 614 in Mentone. The 90-foot-long bridge spans the West Fork of the Little River. Only the center part of the bridge, located on a dirt road that dead-ends on the Old Military Trail, is covered. That 42-foot-long portion was moved to the site in 1972 and set atop an existing bridge.

The original portion of Old Union Crossing Bridge in Mentone is said to have been built in 1863, although this date has not been confirmed. Only the center part of the bridge, located on a dirt road that dead-ends on the Old Military Trail, is covered. That 42-foot-long portion was moved to the site in 1972 and set atop an existing bridge, which is 90 feet long. *Photo by Kelly Kazek.*

The exterior of Old Union Crossing Covered Bridge. *Photo by Kelly Kazek.*

The original portion of Old Union Crossing Bridge is said to have been built in 1863, although there are no records supporting that. According to legend, the bridge got its name during the Civil War when it was built by Union troops over Otter Creek in Lincoln, Alabama. However, the book *Covered Bridges of the United States*, by Warren H. White, states that it was named for the area where it was initially built, called Union Crossing, and locals used it to access New Union Church.

The original covered section was built in the Stringer style. When it was brought to Mentone, it was set atop the existing steel-cable bridge. Motorists can drive across the planked flooring on wheel treads running the length of the bridge. The sides are made of gable board, but they are open for the top six feet so visitors can look into the water. The new bridge is only 10 feet wide, the narrowest in Alabama to carry motor traffic. Because it contains a high percentage of new construction, Old Union Crossing Covered Bridge is considered "non-authentic."

RURAL STUDIO'S UNIQUE
COVERED BRIDGE
AND HOUSE

Thanks to students at Auburn University, Alabama's Black Belt region has some of the most unique architecture in the South. Auburn's Rural Studio was founded in 1993 by D.K. Ruth and Samuel Mockbee to give architecture students hands-on experience while helping people in some of the state's poorest communities build affordable homes and much-needed public buildings such as animal shelters, fire stations, schools and Boys & Girls Clubs. Students in the program build structures from recycled and repurposed materials, working with the philosophy that all people, no matter their socioeconomic status, deserve good design.

The Rural Studio website states, "To most, the measure of success of the Rural Studio is in its built projects; in reality, its success is measured by its effect upon the lives of the students, faculty, families, and communities it touches. It is not only the buildings that make the Rural Studio what it is, but also the education the students receive about architecture and about society. Ultimately, it is about 'sharing the sweat' with the community."

In its first twenty-four years, Rural Studio has built more than 170 projects. Two projects fit into the scope of the topic of this book—one, a covered bridge with an unusual design, and the second, a bridge house. In 2004, students designed and built a contemporary covered bridge using repurposed materials in Perry Lakes Park. The bridge features a tented tin roof. Other projects built by Rural Studio in Perry Lakes Park are a one-hundred-foot-high birding tower, a pavilion and a boardwalk. Students also built three public restrooms, including one that is fifty feet high, that were nominated

This unusual covered bridge in Perry Lakes Park was built by students of Auburn University's Rural Studio. *Courtesy of Billy Milstead/RuralSWAlabama.org*

Students with Auburn University's Rural Studio built this house to span a creek. *Courtesy of Karrie Jacobs.*

in 2015 as finalists in the America's Best Restroom contest sponsored by Cintas, a manufacturer of restroom equipment.

In 2008, the Rural Studio program built one of its $20,000 homes across a creek in Greensboro. It is known as the "bridge house." Projects continue each year at the Rural Studio, and the group also offers floor plans for its $20,000 houses. Learn more at RuralStudio.org.

ALABAMA'S LOST
COVERED BRIDGES

A s the twentieth century progressed, so did building techniques and materials. More and more, heavily trafficked bridges were being built with iron and steel, which did not need protection from the elements. But at least through the 1940s or so, covered bridges continued to be built with some regularity in rural communities and over small waterways.

Gradually, the quaint wooden bridges were replaced, and rather than being saved or moved, most were demolished. Some were destroyed by floods and others by fire, either accidental or arson. Although it would be impossible to know every covered bridge ever built in Alabama, records show structures in at least nineteen counties at some point in history. We compiled a county-by-county list of bridges known to exist, with as much information as possible about each.

A covered bridge in Wilcox County is shown in 1938. It is no longer standing. *Courtesy of Alabama Department of Archives and History.*

AUTAUGA

- Wool Mill, built by slaves of industrialist Daniel Pratt in Prattville over Autauga Creek. It collapsed in 1916.

BARBOUR

- Chattahoochee River Bridge, built in 1833 in Eufaula, this bridge was 540 feet long. It was demolished in 1924.
- Cowikee Creek Bridge, built in Eufaula over Cowikee Creek. The Town Truss–style bridge was demolished in 1913.

BLOUNT

- Blount Springs Bridge, built 1931 over Murphy Creek in Blount Springs.
- Bangor Bridge, built over Mulberry fork in Bangor.
- Chamblee Mill, a 97-foot span over Copeland Creek in Blountsville.
- Clear Springs Bridge, dates unknown.
- Crooked Shoals, built in 1931 over Locust Fork in Nectar.
- Dean's Ferry Bridge was built in 1930 over Locust Fork in the County Line community.
- London Park Bridge, an 11-foot span built in 1979 over a highway culvert near Cleveland.
- Duck Branch, built in 1930.
- Five Points, built over Locust Fork in Blountsville. Swept away in a flood.
- Gable Bridge, a 240-foot span over Mulberry Fork in Blountsville.
- Hayden Bridge, dates unknown.
- Inland Bridge, built in 1930 over Blackburn Fork in Remlap.
- Joy Road Bridge, a 121-foot span over Austin Creek in Blountsville. It was demolished.
- Locust Fork Bridge, built in 1927 over Little Warrior River in Locust Fork. It was replaced in the 1950s.
- Mardis Mill Bridge, built over Graves Creek in Blountsville.

Blount Springs Covered Bridge was built in 1931 over Murphy Creek in Blount Springs. *Courtesy of Blount County Memorial Museum.*

Clear Springs Covered Bridge. *Courtesy of Blount County Memorial Museum.*

Gable Covered Bridge was a 240-foot span over Mulberry Fork in Blountsville. *Courtesy of Blount County Memorial Museum.*

Hayden Covered Bridge in Blount County, dates unknown. *Courtesy of Blount County Memorial Museum.*

- Nectar Bridge, a 385-foot span built in 1934 over Locust Fork in Nectar. It was the seventh-longest bridge in the nation when it burned in 1993. The stone abutments remain at the site.
- Putman Bridge, dates unknown.
- Rockhole Bridge, a 121-foot span over Mulberry Fork in Summit.
- Slab Creek, built in 1933 over Slab Creek in McLarty.
- Snead Bridge, built over Big Mud Creek.
- Standridge Bridge, a 432-foot, Town Truss span built in 1934 over Locust Fork. It was destroyed by arson on November 18, 1967.
- Tyre Green, a 127-foot span built in 1933 over the Little Warrior River in Locust Fork.
- Vaughn, built over Locust Fork in Hayden.
- Ward's Mill, built over Locust Fork in the town of Susan Moore.

CALHOUN

- Cane Creek Bridge, built in 1886 over Cane Creek in Ohatchee. It was destroyed by flood in 1936.
- Chosea Springs, a 98-foot span over Choccolocco Creek in Choccolocco. It was demolished in 1963.
- Hillabee Creek Bridge, built over Hillabee Creek in Hicks.
- Mellon Bridge, a nearly 100-foot span built in the late nineteenth century over Choccolocco Creek in DeArmanville. It was destroyed by arson on October 3, 1970.
- Tallassahatchee Creek Bridge spanned Tallasahatchee Creek.

CHAMBERS

- Double M Farms Bridge, a 75-foot span built over a stream near Lafayette. It was demolished in 2009.

COLBERT

- Big Bear Creek Bridge, built in the mid-nineteenth century in Allsboro.
- Buzzard Roost Bridge, a 94-foot span thought by some historians to have been built circa 1820 over Buzzard Roost Creek in Cherokee. It was one of the first covered bridges built in Alabama. It burned in 1972.
- Cripple Deer Creek Bridge, built in 1859 in Allsboro.

COOSA

- Oakachoy Bridge, a 56-foot span built in 1916 over Tallasseehatchee Creek in Nixburg.

CULLMAN

- Bolte Bridge, built over Brindley Creek in Bolte.
- Cofer Bridge, a 239-foot span built over Ryan Creek in Choccolocco. It was dismantled in 1934.
- Garden City Bridge, a 280-foot span built in 1891 over Mulberry Fork in Garden City. It burned in 1951.

Cofer Covered Bridge in Cullman County is shown in 1933 or 1934. It is no longer standing. *Courtesy of Alabama Department of Archives and History.*

- Gay Bridge, a 123-foot span built in 1898 over Eight Mile Creek in Pleasant Grove, It was dismantled in 1963.
- Kilpatrick Bridge, built in the West Point community. It burned on October 3, 1937.
- Lidy Walker, or Lidy's, Bridge, a 50-foot span built in 1926 over Big Branch in Blount County. In 1958, it was purchased by Lidy Walker for fifty dollars and moved to Lidy's Lake on private property near the city of Cullman. It collapsed in August 2001.
- Mulberry Bridge, a 220-foot span built over Mulberry Fork in Hanceville.
- Putman Bridge, a 474-foot span built over Mulberry Fork in Hanceville.
- Sanford Bridge, a 152-foot span built over Ryan Creek in Bremen.
- Tanner Bridge, a 257-foot span built over Duck Creek in Baileyton.
- Trimble Bridge, a 151-foot span built over Ryan Creek in Trimble.
- Welti Road Bridge, a 100-foot span built circa 1904 over Eight Mile Creek in Welti. It burned on October 22, 1939.
- Whaley Mill Bridge, a 52-foot span built over Whaley Mill Creek in Hanceville.

Putman Covered Bridge, dates unknown. *Courtesy of Blount County Memorial Museum.*

Lidy's Covered Bridge was a 50-foot span built in 1926 in Blount County. *Courtesy of Blount County Memorial Museum.*

ELMORE

- Wetumpka Bridge, built in 1844 over the Coosa River by renowned bridge builder Horace King. It was destroyed by flood in 1886.

ETOWAH

- Duck Springs, a 119-foot span built in 1879 over Big Wills Creek in Duck Springs. It burned in 1972.

LEE

- Chewacla Creek Bridge, thought to be the second covered bridge in this spot in what is now Chewacla State Park, which was developed by the Civilian Conservation Corps in 1939 on the site of Wright's Mill.

Mellon Covered Bridge was a nearly 100-foot span built in the late nineteenth century over Choccolocco Creek in DeArmanville. It was destroyed by arson on October 3, 1970. It is shown here in the 1950s. *Courtesy of Alabama Department of Archives and History.*

A covered bridge over the Chattahoochee River being torn down. *Courtesy of Alabama Department of Archives and History.*

- Loachapoka Bridge, built in 1818 over Sougahatchee Creek in Loachapoka. It was destroyed by flood.
- Meadows Mill Bridge, a 140-foot span built in 1902 over Halawakee Creek in Beulah. The Town Truss–style bridge was located near a historic gristmill. It was destroyed by arson in 1973.
- Woods Bridge, a 100-foot span built over Sougahatchee Creek in Auburn. It collapsed on April 21, 1959.

MADISON

- Butler's Mill Bridge, a 165-foot span built in 1884 over the Paint Rock River near New Hope. It was replaced in the 1940s.

Covered bridge spanning Choccolocco Creek in Old Eastaboga in Talladega County is shown in 1935. *Courtesy of Historic American Buildings Survey/Library of Congress.*

MONTGOMERY

- Norman Bridge, built over Catoma Creek in the city of Montgomery.
- Pintlala Creek Bridge, a 114-foot span built in 1861 over Pintlala Creek in Hope Hull. It was demolished in the mid-twentieth century. It straddled the Montgomery-Lowndes county line on Old Selma Road.
- Wasden Road Bridge, an 82-foot Town Truss–style span built in 1851 over Pintlala Creek by William Trimble. It collapsed in 1965.

RUSSELL

- Dillingham Street Bridge, a 400-foot span built in 1832 over the Chattahoochee River by Horace King. It was destroyed in the Civil War battle of Girard, in what is now Phenix City, in 1865.

SHELBY

- Montevallo Bridge, a 79-foot span built over Shoal Creek in the 1970s.

TALLADEGA

- Lincoln Bridge, a 160-foot span built in 1903 over Choccolocco Creek. It was destroyed by arson in 1963.
- Old Tin Sides, built over Tallaseehatchee Creek in Childersburg.

TALLAPOOSA

- Golden's Mill Bridge, built of Sougahatchee Creek in East Tallassee.
- Hillabee Creek Bridge, built over Hillabee Creek in Alexander City.

Lincoln Bridge, shown here in the 1950s or '60s, was a 160-foot span built in 1903 over Choccolocco Creek. It was destroyed by arson in 1963. *Courtesy of Alabama Department of Archives and History.*

Golden's Mill Covered Bridge was built of Sougahatchee Creek in East Tallassee, date unknown. It is shown here in 1940. *Courtesy of Alabama Department of Archives and History.*

■ Miller Bridge, a 600-foot span built in 1887 over the Tallapoosa River at New Site. At the time, it was the longest covered bridge in the United States. The bridge was built at the launch point of the old Miller's Ferry, which is now part of Horseshoe Bend National Military Park. The bridge was destroyed during a flood in July 1963.

TUSCALOOSA

■ Brookwood Bridge, a 177-foot span built circa 1850 over Hurricane Creek in Brookwood. The Town Truss–style bridge burned in 1965.

GRISTMILLS AND COVERED BRIDGES MAKE PICTURESQUE PAIRS

Covered bridges were often found alongside mills for an obvious reason: mills were operated by running streams that powered the millstones used in grinding cane, corn and various other grains. People on the other side of the water needed a way to get their wagons, and later their cars, across the streams. Add to that the fact that covered wooden bridges were popular at the time when mills were important community hubs, and you have a practical and picturesque combination.

Alabama has at least six mill/covered bridge combinations, although not all are authentic. Some are replicas used to teach people about history, and others were moved to new locations. Here's a look at the history behind the mills and bridges.

KYMULGA GRIST MILL AND COVERED BRIDGE

7346 Grist Mill Road, Childersburg, AL 35044
256-378-7436

Kymulga Grist Mill and Covered Bridge are owned by the City of Childersburg and open to the public as a historic park. The mill was built in 1864 and is one of the few gristmills that survived the Civil War, when most were burned. Three Alabama mills are older: Aderholdt (1836),

Kymulga Grist Mill is a rare four-story mill and also one of the few in Alabama that survived the Civil War. It is located in a public park operated by the City of Childersburg. *Courtesy of Charlotte Kirkpatrick.*

A view of Kymulga Covered Bridge. *Photo by Kelly Kazek.*

Pumpkin Hollow Covered Bridge was built in 1992 in Sterrett over Bear Creek. This 68-foot-long bridge is located in the gated community of New Pumpkin Hollow Lake. The Town Lattice on the bride is decorative and not part of the structure's support. *Courtesy of Josh Box.*

Swan Creek Covered Bridge was built in 2004 on a walking trail in Athens. The 40-foot bridge spans Swan Creek. *Photo by Wil Elrick.*

Hugh King Covered Bridge is 44-foot-long bridge built in 1986 in Springville. *Photo by Wil Elrick.*

The lattice design is visible on Horton Mill Covered Bridge in Oneonta. Rising seventy feet above the Calvert Prong of the Little Warrior River, this is the highest covered bridge in the nation. *Courtesy of Charlotte Kirkpatrick.*

Swann Covered Bridge, also known as the Swann-Joy Covered Bridge, is a 324-foot-long span built in 1933 in Cleveland. *Photo by Wil Elrick.*

Swann Covered Bridge, which spans the Locust Fork of Black Warrior River in Cleveland, is shown in 2010. *Courtesy of Carol M. Highsmith/Library of Congress.*

The historic chapel in Palisades Park in Oneonta. It is one of several structures in a pioneer village in the park, which includes a covered bridge. *Photo by Wil Elrick.*

Price Covered Bridge was built by Claude Price in 2011 in Elberta. *Courtesy of Claude and Virginia Price.*

Horace King Memorial Bridge was built in honor of the master bridge builder in Valley. *Courtesy of Shannon Kazek.*

The covered bridge built by former slave Horace King over the Chattahoochee River. The bridge was demolished in 1924. *Courtesy of Alabama Department of Archives and History.*

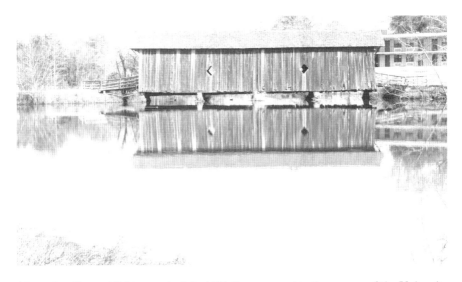

Alamuchee Covered Bridge was built in 1861. It was moved to the campus of the University of West Alabama in 1971. *Courtesy of Charlotte Kirkpatrick.*

Kymulga Covered Bridge is located near Kymulga Grist Mill in a historical park owned by the City of Childersburg. It is open to the public as a historic park. The Kymulga Bridge, built in 1861, is one of only two nineteenth-century bridges that remain in their original locations. Waldo Covered Bridge is the other. The mill was built in 1864 and is one of the few gristmills that survived the Civil War. *Courtesy of Charlotte Kirkpatrick.*

Above: Alamuchee Covered Bridge, built in 1861. *Courtesy of Alabama Department of Archives and History.*

Left: Kymulga Covered Bridge is a 105-foot span built in 1861. *Courtesy of the Alabama Birding Trail.*

The 60-foot-long Poole's Covered Bridge was built in 1998 by Wyndel Eiland to be part of the Pioneer Museum of Alabama in Troy. *Courtesy of Grady Allen.*

Waldo Covered Bridge was built circa 1858 in Talladega, making it the second oldest in Alabama. It is in imminent danger of collapse. It is the only one of the state's historical bridges that is not being maintained. The approaches to this 116-foot-long bridge are missing, so there is no access. *Courtesy of Phillip M. Burrow.*

Pea Ridge Covered Bridge is shown in the 1950s or '60s. *Courtesy of Alabama Department of Archives and History.*

A covered bridge in Talladega National Forest near Oxford is shown in 1959. *Courtesy of Alabama Department of Archives and History.*

This unusual covered bridge in Perry Lakes Park was built by students of Auburn University's Rural Studio. *Courtesy of Paul Franklin/Alabama Birding Trail.*

The entrance of Easley Covered Bridge. *Photo by Wil Elrick.*

Left: View of Town Truss construction inside Salem-Shotwell Covered Bridge. *Courtesy of Shannon Kazek.*

Below: Gilbert's Covered Bridge was built in 1999 by the Gilbert family in Limestone County. *Courtesy of Don Osborne.*

Ivalee Covered Bridge in Attalla. *Photo by Kelly Kazek.*

Old Union Crossing Covered Bridge in Mentone. *Courtesy of Charlotte Kirkpatrick.*

Overland Road Covered Bridge in Brierfield Ironworks Park in Bibb County. *Photo by Kelly Kazek.*

One of two green-and-white covered bridges spans a pond in Governors Park located directly behind Holly Pond Town Hall. *Photo by Wil Elrick.*

Swan Creek Covered Bridge on the walking trail in Athens. *Photo by Wil Elrick.*

This bridge, shown in 2010, was built in 1972 in the Tannehill Valley Estates subdivision. *Courtesy of the George F. Landegger Collection of Alabama Photographs/Carol M. Highsmith's America/ Library of Congress.*

Gargus Covered Bridge was built in 1966 in Gallant. *Photo by Wil Elrick.*

Pumpkin Hollow Covered Bridge in Sterrett. *Courtesy of Josh Box.*

Cambron Covered Bridge was built in 1974 on the Green Mountain Nature Trail in Huntsville. *Photo by Wil Elrick.*

Twin Creeks Covered Bridge near Midway is the longest in the state at 334 feet. *Courtesy of John Trent.*

Riddle's (1844) and Rikard's (circa 1850s). Aderholdt does not have a bridge at the site.

Kymulga is the most unusual mill in the state because it is four stories tall and was run by turbines beneath the mill rather than the typical water wheel.

It was built for Confederate lieutenant colonel George H. Forney, but he died at the Battle of Wilderness that May before it was completed. His widow oversaw completion of the mill before selling it. Historians believe that the decorative trim beneath the eaves, unusual for such a utilitarian structure, was added by Mrs. Forney.

The mill has three stories in the front and four on the back, where it runs down to the waterline. Three underwater turbines power the mill's lights and grain elevator. Two of Kymulga's five sets of millstones are French buhrs, known as the hardest rock in the world. Grain is still ground at the mill, and samples can be purchased in the gift shop. Because of its size, age and location on the water, the mill is in need of constant maintenance. A three-dollar entry fee to the park helps fund repairs, but park administrators also rely on private donations. You can donate to the mill by visiting the website at kymulgagristmill.com or calling 256-378-7436.

Kymulga Covered Bridge was built in 1861 over Talladega Creek. The bridge, of Howe Truss construction, is 105 feet long.

RIKARD'S MILL AND COVERED BRIDGE

4116 Alabama Highway 265 North, Beatrice, AL 36425
251-789-2781

Rikard's Mill Historical Park, owned by the Monroe County Heritage Museums, features a mule-driven syrup mill, a replica cabin, a barn exhibit, a blacksmith shop and the covered bridge gift shop. Located on Flat Creek near the town of Beatrice, the circa late 1850s mill was operated by the Rikard family until the 1960s. The first mill was built on the property by Jake Rikard in 1845 but was soon destroyed by a flood. The replacement bridge was restored in 1993 and donated by descendants of Jake Rikard to the Heritage Museums. It is listed in the Alabama Register of Landmarks and Heritage.

Rikard's Mill was constructed near the town of Beatrice by Jacob Rikard, a local blacksmith. It was operated by the Rikard family until the 1960s. The family eventually donated the mill to the Monroe County Heritage Museums. The mill was placed on the Alabama Register of Landmarks and Heritage in 1998. *Courtesy of Billy Milstead/ RuralSWAlabama.org.*

Rikard's Mill Covered Bridge, an 81-foot-long span, is one of the most unusual in Alabama because it's actually a gift shop for Rikard's Mill Park, which includes a circa 1840s gristmill, syrup mill, blacksmith shop and replica pioneer cabin. The bridge was constructed atop an old steel bridge in 1994 with a shop built on top. *Courtesy of Billy Milstead/RuralSWAlabama.org.*

The "covered" bridge is really a small building constructed over a stream where visitors can buy souvenirs. Although the mill still functions, both to grind grain and make syrup, the park's shops were closed in 2017, so be sure to call before going.

RIDDLE CANE MILL AND COVERED BRIDGE

57900 Alabama Highway 77, Talladega, AL 35160
256-761-2050

Riddle's Mill in Waldo was built around the 1850s along the banks of Talladega Creek by the Riddle brothers. The Riddle family built a foundry nearby, called Maria's Forge after the family patriarch and wife of Samuel Riddle, Maria Waldo Riddle. The mill was used for political rallies, dances, religious services and other community events. It was even used as a coffin shop, where wooden caskets were built. The mill was powered by an iron turbine in a water box rather than the conventional water wheel.

The mill and bridge are near Riddle's Hole, an 1840 gold mine that continued operating until World War II. It was one of about seven gold

Riddle's Mill, an abandoned gristmill that was later made into the town hall for the tiny burg of Waldo and then a restaurant. The mill is currently empty. *Photo by Charles Byrom.*

Waldo Covered Bridge was built circa 1858 in Talladega, making it the second oldest in Alabama. It is in imminent danger of collapse. It is the only one of the state's historical bridges that is not being maintained. The approaches to this 116-foot-long bridge are missing, so there is no access. *Courtesy of Glenn Wills.*

mines in the area. Since mill operations stopped, the building was moved a short distance from its original location and used as the Waldo Town Hall and a restaurant. It is currently empty.

The nearby bridge, referred to as both Riddle's Mill Bridge and Waldo Bridge, spans Talladega Creek in the Waldo community. It was built circa 1858. According to legend, the bridge was used by Wilson's Raiders during the Civil War. The bridge, which spans 115 feet, no longer has approaches and is not maintained.

CLARKSON COVERED BRIDGE AND MILL

1240 County Road 1043, Cullman, AL 35057
256-739-2916

Clarkson Covered Bridge, also called Clarkson-Legg Covered Bridge, was built in 1904 in Cullman. At 270 feet long, it is the third longest in Alabama,

Clarkson Covered Bridge Clarkson, also known as Clarkson-Legg Covered Bridge, is a 250-foot-long span over Crooked Creek in Cullman. It was built in 1904 and is now part of a park that includes a working gristmill and dogtrot cabin. *Photo by Wil Elrick.*

A replica gristmill was built as part of the historical park beside Clarkson Covered Bridge in Cullman. *Courtesy of Charlotte Kirkpatrick.*

after Twin Creeks and Swann Bridges. Visitors can walk across it, although it is not viable for motor traffic.

The Clarkson bridge was torn into two pieces by a storm in 1921, but the segments were salvaged and restored. The bridge was restored again in 1975 for America's bicentennial. As part of the bicentennial project, the county added hiking trails, cabins, a picnic area and two period structures to create a historical park. The dogtrot log cabin and working gristmill were designed to help accent the bridge and local history. The bridge and park are often used for weddings, car shows and other community activities.

PIONEER MUSEUM OF ALABAMA

248 U.S. Highway 231 North, Troy, AL 36081
334-566-3597

In 1971, Curren and Margaret Farmer of Troy founded the Pioneer Museum of Alabama with a goal of teaching people about Alabama's past. It began with a single log cabin but now has twenty-two historical structures on forty acres, as well as twenty-five thousand artifacts.

Visitors to the site can see an 1800s steam locomotive, a copper turpentine still, a corncrib, a nineteenth-century church, log cabins, a horse-drawn jail (aka paddy wagon) and a tree grown from seeds taken to the moon in a 1971 Apollo mission. The museum also features a working gristmill and a rustic covered bridge. Called Poole's Bridge, the 60-foot-long span was constructed in 1998 as part of the museum.

The site is open for tours as well as interactive folk life demonstrations. Annual events include the Pioneer Days, Spring Plantin' and Old Time Christmas. Pioneer Days is a weekend event held each fall featuring demonstrations of blacksmithing, woodworking, rope-making, spinning and quilting.

The 60-foot-long Poole's Covered Bridge was built in 1998 by Wyndel Eiland to be part of the Pioneer Museum of Alabama in Troy. *Courtesy of Pamela Watts Cooper.*

NOCCALULA FALLS STATE PARK

1500 Noccalula Road, Gadsden, AL 35904
256-549-4663

Gilliland's Bridge is an 85-foot-long bridge built in 1899 in Reece City. It was moved in 1968 to Noccalula Falls State Park, which features a pioneer village. The bridge crosses a pond. A gristmill replica was built near the bridge, along with other pioneer buildings that include a log home, a smokehouse, a blacksmith shop and a post office.

Right: This gristmill is located near Gilliland Covered Bridge in Noccalula State Park. *Photo by Kelly Kazek.*

Below: Noccalula Falls drop ninety feet into Black Creek Gorge. The falls, shown here behind a statue of the legendary Princess Noccalula, are the centerpiece of Noccalula Falls State Park in Gadsden. *Photo by Kelly Kazek.*

The park's centerpiece is Noccalula Falls, which drops ninety feet from a ledge that is part of Lookout Mountain. The falls can be viewed at no charge, but admission is charged to visit the rest of the park, which also includes a botanical garden, kiddie train, camping, picnic areas and more.

TYPES OF
COVERED BRIDGES

Today, people love covered bridges because they are beautiful and unusual, but their original purpose was to provide a practical way to cross obstructions.

The Federal Highway Administration's *Covered Bridge Manual* defines an authentic bridge as one that "uses heavy timber trusses to carry loads over an obstruction. The floor system spans between the longitudinal trusses and distributes and carries the loads between those trusses. The bridge is completed by lateral bracing (elements that connect each truss, or side, of the bridge), a wall system, and roof intended to prevent weathering. The roof's primary function is to protect the structural timbers from the ravages of intermittent wetting. In bridge engineering terms, this style of structure is termed a through truss."

The flooring system spans the trusses, distributing and carrying the loads in between. The bridge is completed with some type of lateral bracing that connects to the trusses for support. The trusses are built to rest on, or connect to, abutments and piers, which give them height to clear waterways and their banks. The walls and the roof are the final additions to the bridge and, more often than not, are also connected to the lateral bracing for added support.

The true bones of the covered bridges are the truss system, but there are a variety of styles. Here are the most common types of covered bridges.

KINGPOST/KING TRUSS TRUSS

The Kingpost Truss design, which looks like a simple triangle, is the oldest and simplest of truss designs. It also influenced many truss styles to follow.

The inclined timbers (or the triangle sides) serve as both the top chord and diagonals and resist the compression forces on the bridge. The horizontal timber (the bottom of the triangle) acts as the bottom chord and provides tension for the bridge.

A central vertical timber known as the kingpost (a line that goes from the tip of the triangle to the center of the bottom) connects the other timbers of the truss and provides additional tension for the bridge. The Kingpost Truss is one of the simplest designs, yet it is quite strong. This type of design is used mostly on spans of 30 feet or less.

The only remaining Kingpost Truss bridge remaining in Alabama is Jimmy Lay Covered Bridge in Walker County.

QUEENPOST TRUSS

The Queenpost Truss design is a modified version of the Kingpost Truss but for use on longer spans. The Queenpost is best described as a Kingpost Truss that is stretched to include a center rectangular section, thereby expanding the length of the bridge. The classic examples of the Queenpost-style bridges do not have any diagonal timbers in the center section, so technically the design is not a truss at all but a frame. Some Queenpost Trusses have additional counterbraces for supporting the angles on each side of the bridge. These bridges usually span lengths of 40 to 75 feet. Some examples of Queenpost Truss covered bridges remaining in Alabama are Price Covered Bridge in Baldwin County and Waldo (or Riddle Mill) Covered Bridge in Talladega County, which is a combination of Howe Truss and Queenpost Truss design.

MULTIPLE KINGPOST TRUSS

The Multiple Kingpost Truss features a center Kingpost Truss with a series of right-angle panels that connect to a horizontal top timber. This results

Jimmy Lay Covered Bridge in Empire. *Courtesy of Wil Elrick.*

The interior of Coldwater Covered Bridge in Oxford. It is a 63-foot span built no later than 1850, and probably earlier, making it the oldest covered bridge in Alabama. *Photo by Wil Elrick.*

in multiple vertical posts stretching between the top and bottom timbers, with one diagonal timber for each vertical one. This creates multiple short sections along the truss span. The Multiple Kingpost Truss is designed for spans ranging from 36 to 124 feet. The only recorded bridge with this design remaining in Alabama is Coldwater Covered Bridge in Calhoun County.

BURR/BURR ARCH TRUSS

The Burr Arch Truss is best described as two long arches resting on abutments on each side of an obstacle, such as a river, with a Multiple Kingpost Truss sandwiched between them. The addition of the arch to the Multiple Kingpost Truss provided more strength and stability, allowing the bridges to span greater lengths and carry heavier loads. The design for the Burr Arch Truss was issued the first-ever United States patent for a bridge timber-truss system. It was issued to Theodore Burr in 1806 and became immediately popular with bridge builders due to its strength and durability. Burr had used the truss system when he built a non-covered bridge from Waterford to Lansingburgh over the Hudson River in Waterford, New York, in 1804. More covered bridges in the United States are built with the Burr Arch Truss than any other type, the *Covered Bridge Manual* notes. It can be used to span lengths greater than 220 feet.

Burr's Waterford-to-Lansingburgh Bridge, or Union Bridge, was known as the greatest wooden span of its time. Fittingly, this bridge, which was the blueprint for so many of the covered bridges constructed in the United States, was eventually covered to protect the trusses from the elements. It was in use for 105 years until 1909, when it was destroyed in a gas fire.

One interesting historical note to Theodore Burr: While Burr was prominently known as one of the United States' bridge pioneers, he didn't have as much notoriety as his cousin, U.S. Vice President Aaron Burr, who shot and killed Alexander Hamilton in a duel in Weehawken, New Jersey, on July 11, 1804, the same year Theodore was building a bridge just upriver from the site of the duel. After the duel and the public outcry that followed, Aaron Burr left politics, only to become entangled in an alleged plot to overthrow the federal government. The charges of treason against Burr made him a wanted fugitive from the United States government. Burr fled south toward Spanish-controlled Florida, but he was eventually captured in the community of Wakefield, Alabama, near the Tombigbee River. At the time of Burr's arrest, Wakefield was part of the Mississippi Territory but is now part of the state of Alabama.

There are no recorded Burr, or Burr Arch, Truss covered bridges remaining in Alabama.

TOWN TRUSS AND TOWN LATTICE TRUSS

Ithiel Town, born in Connecticut in 1784, patented a design to stabilize the sides of covered bridges that became popular in New England and the South. Called the Town Lattice Truss, the design was patented in 1820.

Town said that he came up with the design as a response to this question: "By what construction or arrangement will the least quantity of materials, and cost of labor, erect a bridge of any practicable span or opening between piers or abutments, to be the strongest and most permanent, and to admit of the easiest repair?"

The solution, according to an article by John L. Sanders, was

> *to build a continuous timber frame box, consisting of a set of vertical bridge walls...composed of continuous timber latticework. These lattice walls consisted of grids of closely set, diagonally intersecting sawn plank, 10 to 11 inches wide and 3 to 3.5 inches thick, pinned together at each crossing by three or four wooden "trunnels" or tree nails, and similarly pinned where the diagonals overlapped the horizontal lengthwise string pieces at the bottom and top of each wall. Horizontal beams and rafters bound the lattice walls into a rigid framework, which carried a plank floor and above, a shingled roof. Weatherboarding protected the structure from the weather.*

The Town Lattice Truss is so named because the resulting crisscross pattern looks like latticework. This bridge is different from the other bridge styles, which rely on very heavy timbers and elaborate joint connections to create the support for the bridge. They were often expensive and required skilled artisans to build them. The Town Lattice design was the opposite, being made entirely of planks that could be easily managed by unskilled workers and built by individual craftsmen. Many bridge aficionados argue that the Town Lattice Truss was the single most important development in the history of covered bridges. It is the style used on many private covered bridges and could be used in spans ranging from 25 to about 165 feet.

This interior photo of Horton Mill Covered bridge shows the Town Lattice Truss construction. *Photo by Wil Elrick.*

The Town Lattice Truss construction inside Easley Covered Bridge. *Photo by Wil Elrick.*

The exterior of Horton Mill Covered Bridge. *Photo by Wil Elrick.*

The Thomas Bibb House in Huntsville has Town Lattice Truss supports beneath its roof. It is shown here in 1934. *Courtesy of Historic American Buildings Survey/Library of Congress.*

Town's patented design eventually made him a wealthy man. He built numerous covered bridges himself, particularly in North Carolina, but other builders paid to use his design, which can still be seen in several covered bridges in Alabama.

Interestingly, the lattice truss design was occasionally used for roof support in homes and buildings. One of the few surviving examples is visible in the attic of the Thomas Bibb Home in Huntsville, Alabama, according to an article by local historian Eleanor Newman Hutchens. The Bibb house was built in 1836 at 303 Williams Street.

Some examples of the Town Lattice Truss covered bridges remaining in Alabama are Alamuchee in Sumter County, Clarkson in Cullman, Easley in Blount County, Horace King Memorial Bridge in Valley, Horton Mill in Blount County, Salem-Shotwell in Lee County and Swann in Blount County.

LONG TRUSS BRIDGE

The Long Truss looks like a series of connected diagonal cross beams in each section, with a twist. The design features top and bottom horizontal beams made of heavy timers with vertical supports in each section, also made by heavy timbers. Each section then has two diagonal beams in different directions and crossing in the middle. The twist on this style is that where the diagonal supports connect to the upper and lower connections, there are wedges that allow builders and maintenance workers to adjust the compression at each joint. The covered versions of them are good for spanning obstacles from 50 to 170 feet in length.

Saunders Covered Bridge is a 51-foot-long bridge over a spillway of Lake Laurlee. *Courtesy of Chris McCune and Ben Richardson.*

The Long Truss configuration was patented by U.S. Army engineer Colonel Stephen H. Long in 1830. Rather than relying on empirical design, this style was created from engineering analysis and introduced the groundbreaking engineering concept of prestressing. The Long Truss design would be used in the construction of highway and railway bridges for decades in the eastern United States.

An example of a Long Truss covered bridge remaining in Alabama is Bob Saunders Family Covered Bridge in Shelby County.

HOWE TRUSS BRIDGE

The Howe Truss is similar in design to the Long Truss, consisting of horizontal top and bottom beams with vertical supports with crossing diagonal supports in each section. The major difference between the Long Truss and the Howe Truss was the vertical supports, which were made of metal rods in the Howe design rather than wooden beams. Threaded connections on the ends of the metal rods made adjusting the tension on the bridge relatively easy for builders and maintainers. The design of this truss made it adaptable to various situations and good for spanning obstacles from 20 feet in length up to 200 feet.

The initial patent for the Howe Truss was issued to architect William Howe in 1840. The design was then purchased by Amasa Stone for exclusive use in New England. Six years later, Howe submitted a second patent with several small improvements in 1846 as the new Howe Truss design. The design overtook the popular Long Truss because of the less expensive and easier-to-maintain metal rods. The Howe Truss was the first patent for the truss design to feature major structural components made from metal.

The Howe Truss was used predominantly in the design of railroad bridges in the western parts of the country and was second only to the Burr Arch Truss in popularity. Howe Truss bridges are still built by aficionados today. In Alabama, a surviving example is Kymulga Covered Bridge in Talladega County. Waldo Covered Bridge is considered a Howe-Queen combination.

OTHER TYPES OF TRUSSES

The previous seven types of truss bridges are the most common styles used in the United States, as identified by the Federal Highway Administration. But there are other types of truss designs you might see across the country, including Brown, Childs, Haupt, Inverted Bowstring, Paddleford, Paddleford Arch, Partridge, Post, Pratt, Smith, Tied Arch, Supplemental Town, Double Town, Warren and Double Warren. The FHA has also identified twenty-seven unnamed types of covered bridges in the United States.

STRINGER, OR BEAM, COVERED BRIDGES

Stringer style is the simplest form of bridge design; it does not involve trusses. Construction features a series of longitudinal aligned beams, or stringers, placed beneath the decking beams, running parallel to the bridge. The stringers are placed close together to support the decking without using a truss system. This design is commonly used in small bridges on rural roads. Today, stringers are often made from steel or concrete, with concrete decking. These bridges can be built in any length up to about 250 feet. To span lengths of more than 250 feet, multiple stringer bridges are joined together creating what is known as a continuous span.

The Stringer style may be the most maligned design in the fascinating history of covered bridges. Many historians consider bridges built with stringers instead of traditional-style trusses as "non-authentic" because they are coverings built on top of bridges that play no role in supporting the structure.

However, some forms of Stringer-style bridges are considered part of a truss system. There are also covered bridges that have been saved and repaired using a Stringer-type system, which can be considered authentic and non-authentic at the same time, much like Schrödinger's cat can be considered both alive and dead at the same time.

The majority of Alabama's covered bridges are built in the Stringer style. Examples are Amanda's in Etowah County, Askew in Lee County, Blount County Master Gardeners Memorial, Clearbranch United Methodist Church in Jefferson County, Cooley in Jefferson County, Creekwood in Limestone County, Double M Farms in Chambers County, First Baptist Church in Talladega County, Foster's in Houston County, Gabe's in Houston

County, Gargus in Etowah County, Gibson in Cleburne County, Gilbert in Limestone County, Gilliland (or Reese) in Etowah County, Governors Park No. 1 and No. 2 in Cullman County, Hugh King in St. Clair County, Hooters in Houston County, Indian in DeKalb County, Ivalee in Etowah County, Jeff Barton in Walker County, Marriott Grand Hotel Point Clear Resort and Spa, Mason in Blount County, Mother's in Elmore County, Old Downing Mill in Calhoun County, Old Union Crossing in DeKalb County, Overland Road in Bibb County, Perry Lakes Park in Perry County, Poole's in Pike County, Richard Martin Trail in Limestone County, Romine in Limestone County, Sandagger in Mobile County, Swan Creek Walking Trail in Limestone County, Tannehill Valley Estates in Jefferson County, Twin Creeks in Bullock County and Valley Creek Park in Dallas County.

RECORD-BREAKING BRIDGES
AND OTHER TRIVIA

A labama is home to the nation's highest covered bridge, according to the National Park Service. At one time, the state also had bragging rights to the nation's longest covered bridge, Miller Bridge in Tallapoosa County, before it washed away in a 1963 flood. Here is more trivia about covered bridges:

- Covered bridges can be found in at least twenty-nine states.

- According to the *World Guide to Covered Bridges*, more than half of the world's covered bridges are located in the United States—about 880 of the estimated 1,400 bridges worldwide.

- Only certain bridges are considered "authentic." The Federal Highway Administration defines authentic bridges as those built with "heavy timber trusses to carry loads over an obstruction."

- The highest covered bridge in the nation is Horton Mill in Oneonta, Alabama, which is seventy feet above the Calvert Prong of the Little Warrior River.

- The longest covered authentic, historical bridge in Alabama is the 324-foot Swann Covered Bridge in Cleveland. The longest overall is Twin Creeks Covered Bridge, built in 2000 at Wehle Land

Coldwater Covered Bridge in Oxford is a 63-foot span built no later than 1850, and probably earlier, making it the oldest covered bridge in Alabama. *Photo by Wil Elrick.*

Swann Covered Bridge spanning the Locust Fork of the Black Warrior River is shown in the 1960s or '70s. *Courtesy of the Alabama Department of Archives and History.*

Twin Creeks Covered Bridge near Midway is the longest in the state at 334 feet. *Courtesy of John Trent.*

Conservation Center near Midway. It is 334 feet long and built over a concrete base rather than water. Nectar Covered Bridge near Cleveland was the longest in Alabama at 385 feet until it burned in 1993.

- The longest bridge ever built in Alabama was the 600-foot long Miller Bridge over the Tallapoosa River near Alexander City. It was washed away in a flood in 1963. The site where it stood is part of Horseshoe Bend National Military Park.

- The longest covered bridge in the world is the Hartland Bridge, a 1,282-foot span in New Brunswick, Canada.

- The longest covered bridge in the United States as of April 2018 is the 613-foot Smolen-Gulf Bridge, which was built in 2008 in Ashtabula County, Ohio.

- Alabama's oldest span is Coldwater Covered Bridge, which was built in 1850 in Oxford.

- It is unclear which bridge is the oldest in the United States, but three bridges were reportedly built before 1830: Hyde Hall Bridge in Oswego County, New York, circa 1825; Haverhill-Bath Bridge at Woodsville, New Hampshire, circa 1829; and the Roberts Bridge in Preble County, Ohio, built in 1829.

- The earliest known covered bridge in the United States was the 550-foot Permanent Bridge, which was built in 1805 over the Schuylkill River in Philadelphia, according to the Covered Bridge Society. It was destroyed by fire in 1875.

- Alabama has at least fifty-one existing covered bridges. The state with most covered bridges is Pennsylvania with 213, followed by Ohio with 148.

- Six of Alabama's bridges are in the National Register of Historic Places: Clarkson, Coldwater, Easley, Horton Mill, Kymulga and Swann. Nectar was also listed in the National Register before it was lost to arson in 1993.

- The nation's oldest bridges were built in the Middle Atlantic and New England states. However, rather than following the westward progression of settlements, bridge building bypassed intervening states and jumped to California in 1862, according to the Federal Highway Administration's *Covered Bridge Manual.* "This may well reflect the leapfrog of population that followed the California Gold Rush, as well as the availability of large timbers in the West," the publication notes.

- The *Covered Bridge Manual* gives a breakdown of surviving bridges built by decade: 195 bridges built between 1870 and 1879 survive, and 149 built between 1880 and 1889. Building continued to decline, and only 7 bridges survive that were built between 1940 and 1949 and only 5 from 1950 to 1959. "More than a dozen were built in each decade of the 1960s, 1970s, and 1980s, which shows a recent resurgence in their popularity," according to the manual.

TALES OF HAUNTINGS AT ALABAMA'S COVERED BRIDGES

On a dark and stormy night, what could be spookier than a shadowy covered bridge spanning the banks of a river filled with rushing and crashing water? With darkness encompassing you, timbers creaking, water rushing below and lightning crackling, you're transported from the idyllic covered bridge of daytime to a slowly strangling nightmare. The image of a specter bolting from beneath a covered bridge has haunted us since Ichabod Crane was chased by the Headless Horseman in his infamous ride in Washington Irving's classic "The Legend of Sleepy Hollow."

The same characteristic that makes covered bridges seem romantic makes them ripe for ghostly campfire tales at night. While Alabama is far from the quaint village of Sleepy Hollow, its covered bridges are home to several legends of horrible haints and ghastly ghosts.

ALAMUCHEE COVERED BRIDGE

Locals in the small town of Livingston have talked about a ghost at the Alamuchee Covered Bridge for more than 130 years. It is said to be the ghost of Alabama's Outlaw Sheriff Steve Renfroe, described earlier, who was hanged at the bridge by a lynch mob to stop of his reign of terror that held the town hostage for many months.

Renfroe was hanged on the night of July 13, 1886, and one version of the story says that each year, on the anniversary of his death, he returns to the bridge

Alamuchee Covered Bridge is reportedly haunted. *Courtesy of Charlotte Kirkpatrick.*

and makes his presence known. Another version of the story says that the outlaw haunts the bridge nightly, trying to attract the attention of anyone passing the bridge. A final version of the story says that he was hanged from a chinaberry tree growing on the riverbank at the bridge's end and that the tree's limbs still shake in the breeze as if Renfroe's body still convulses there. Even though the bridge was moved from its original location spanning the Sucarnoochee River to the University of West Alabama campus, old-timers still refuse to cross the bridge for fear the ghost had moved with the structure.

SALEM-SHOTWELL COVERED BRIDGE

Legends of ghosts that haunt the Salem-Shotwell Covered Bridge date back more than one hundred years. The original span was built in 1900, but after it was swept away in a 2005 storm, it was rebuilt from the recovered parts in a new location, Opelika Municipal Park.

In the book *Haunted Auburn and Opelika*, the authors recall tales of Native American spirits reaching up from the waters of Wacoochee Creek to visitors passing on the covered bridge. Some tales even suggest that these spirits attempted to grab people and pull them into the water to their deaths.

While native spirits are by far the oldest subjects of these tales, Salem-Shotwell Bridge is most commonly connected to the tale of a woman and her two children who were killed in a car accident on the bridge and now haunt the bridge. Legend says that the children's ghosts will appear if visitors leave candy on the bridge. Another spirit is said to be that of a young woman who was strangled to death on the bridge.

Here is a story posted on ghostvillage.com by a woman named Christine recounting a ghost-hunting trip to Salem-Shotwell bridge:

> *You are gonna flip your lid when I tell you this! First of all, I just came from the Shotwell Covered Bridge. Don't know if you know, but two kids died in a car accident and a woman got killed and raped. So, we put some candy there to see if the kids would come and take it like they say. We get out of the car and we see these two dogs at the top of the road, about one hundred yards from us. They were creeping me out 'cause they were just sitting there watching us. We walked to the other side of the bridge and back to the middle, where my friend laid down the candy. As soon as he laid down the candy, I snapped a picture and the dogs went wild, barking and growling. It scared us, so we ran to the car. As we approached the car, the dogs ran towards us and stopped dead in their tracks and high tailed it back the way they came. Maybe they sensed something. It was like they were trying to warn us.*

This ghost hunt took place before the bridge was destroyed by a storm on June 5, 2005.

Other ghostly legends attached to this bridge include the following tale. In the 1960s, a young woman was reportedly choked to death or hanged herself on this bridge. One version says that she invited a young man to meet her on the bridge for a late-night tryst, and he strangled her. The other version says that the young man did not show up, and the young woman was so upset that she hanged herself from the bridge's rafters. Yet another version has the dejected young lady waiting for her prom date who never shows and then hanging herself from the rafters in her prom dress.

Another legend recounts the death of a young woman who was driving in a storm one night. As she approached the bridge, she lost control as the car skidded in the curve of the wet road. She crashed over the edge into the storm-riled waters below. It is said that her ghost floats in the waters of Wacoochee Creek, accompanied by the smell of her burning flesh, presumably from a fire caused by the crash.

Salem-Shotwell Covered Bridge is shown in a photo taken in the 1960s or '70s. *Courtesy of Alabama Department of Archives and History.*

One other ghostly tale involving the Salem-Shotwell Covered Bridge is connected only to its current location in Opelika Municipal Park. It involves the ghost of a young boy that is barefoot and plays in the creek, either on the bridge or in the nearby playground. There is no known backstory for this spirit, but apparently, only children can see him. On occasion, adults have claimed to hear the child's ghostly voice calling to their children, inviting them to "come play with me."

One interesting fact adds to the ghostly lore involving this bridge: the Lee County Commission approved the transfer of ownership of the remains of the Salem-Shotwell Covered Bridge to the City of Opelika in a meeting that was held on October 31, 2005.

OAKACHOY COVERED BRIDGE

Oakachoy Covered Bridge, also known as Thomas Covered Bridge, spanned Oakachoy Creek in Coosa County from 1916 until 2001. It was said by locals to have been haunted throughout its existence. While the bridge was standing, motorists were said to experience such things as engines dying, door handles shaking or vehicles being rocked while stopped on the bridge. Since the bridge was destroyed by arson in 2001, ghostly phenomena are still reported at the site. One story centers on an apparition, sometimes a black shadowy form walking near the woods, that disappears as suddenly as it appears.

NECTAR COVERED BRIDGE

Two stone pillars standing in the gently rolling water of the Locust Fork branch of the Black Warrior River are a ghostly reminder of what was Alabama's longest covered bridge, known as the Nectar Covered Bridge. Standing above the muddy waters from 1932 until it was destroyed by arson in 1993, the bridge was a place for community gatherings, baptisms and hauntings. The bridge was said to be haunted by a mailman who was supposedly killed inside the walls of the 385-foot span, although no evidence could be found to support this claim. The modern postal service does not have any records of mailmen dying on Nectar Bridge, and there are no supporting documents to show anyone ever died on the historic structure. Tales of the ghostly delivery man continuing his ethereal duties persist to this day.

After the bridge's demise, the road was routed a few hundred feet north, onto a new concrete bridge, but vestiges of the old roadway are still visible. Walking down the path to the river, visitors can get close to the remaining giant pillars and easily trace the path the original road took. At dusk, it is easy to imagine a ghostly figure standing at the edge of the bridge abutment, trying to complete his route.

Nectar Covered Bridge, built in 1934 in Oneonta, was lost to arson in 1993. *Photo by Wil Elrick.*

CRY BABY BRIDGES

If you read the earlier section on Salem-Shotwell Covered Bridge and thought that parts of it seemed familiar, it's because you've likely heard similar tales. It had all the elements of common urban legend about drowned children and a desperate mother. This legend has long been connected to bridges and is known as the "Cry Baby Bridge." It seems that every state, if not every community across the country, has a mysterious location that locals refer to as "Cry Baby Bridge," "Cry Baby Creek" or, in many instances, "Cry Baby Holler." Alabama is no different. Following are a few versions of these stories that make the rounds.

The most common legend starts with a one-lane bridge in a rural area. Reportedly, motorists who stop on the bridge at night will hear a baby crying. According to lore, the baby was drowned in the creek by its mother, who went insane and thought that the baby was evil.

One disturbing legend says that the natives who inhabited the area long ago would take their sick infants to a creek and drown them or leave them to die. Now, the story goes, people can hear the crying of lost infants if they stop on the bridge at night.

Another tale says that those who leave a candy bar on the edge of the bridge and leave will return to find that a bite has been taken out of it. They may also hear an infant crying or a woman sobbing. A story behind this legend says that in the 1800s, a wagon crossing a wooden bridge crashed, sending a young mother and her infant child into the creek below. The infant drowned, but the mother survived. Mysterious sounds are attributed to the cries of the baby and the weeping of the grief-stricken mother.

Here's another tale: If you go to the bridge and stop your car briefly before driving away, you will later find tiny handprints from a ghostly child covering your vehicle. The child in this tale allegedly died in the 1950s or 1960s in a car accident on the bridge.

Variants of this urban legend can be heard about bridges throughout Alabama and across the United States. A quick Internet search turns up well over one hundred rumored bridges with this legend attached. The bridges seem to have several things in common:

- They are one-lane bridges over creeks or streams.
- They are in rural areas.
- They attract people who want to experiment with the legends, or "ghost hunt," including some who leave behind graffiti.

The view from Old Union Crossing Covered Bridge in Mentone. *Photo by Kelly Kazek.*

Two similar urban legends involve bridges but don't include babies. If you go to a reportedly haunted bridge, turn off your headlights and put your car in neutral, legend says that it will be pushed across the bridge by a young woman who was killed by a car while walking along the bridge. She reportedly pushes your car across the bridge to keep you from the same fate. Another version of the legend says that those who stop their cars on the bridge will see ghostly headlights appear from nowhere and chase the car from the bridge. The apparition reportedly is driven by the ghost of a person killed in a car in the first part of the twentieth century, and the spirit doesn't want others to meet the same fate.

None of the legends names specific people, making researching their origins all but impossible. Most researchers theorize that they are cautionary tales meant to warn people of potentially dangerous bridges or surroundings. Whatever their origins, stories such as these predate the invention of the automobile. But they show no signs of fading—it's a safe bet that most any young person could direct you to a bridge with such a legend attached. In fact, it's likely that whoever you ask will be glad to share their own experience of the haunted bridge and the forever crying baby.

BRIDGES DOCUMENTED BY THE HISTORIC AMERICAN BUILDINGS SURVEY

Four of Alabama's historic covered bridges were documented in the 1930s by the Historic American Buildings Survey (HABS). The survey, a division of the National Park Service, was created in 1933 as the first federal preservation program. It was one of numerous programs under Franklin D. Roosevelt's New Deal meant to create jobs for people who were out of work during the Depression—in this case architects, draftsmen and photographers.

HABS was meant to document the country's architectural heritage and "mitigate the negative effects upon our history and culture of rapidly vanishing architectural resources," the National Park Service noted. Numerous historic homes in Alabama were also catalogued in the 1930s, including interior and exterior photos as well as blueprints, providing invaluable documentation of structures that have since collapsed or been destroyed.

Along with the homes and other buildings, HABS documented "more than 100 covered bridges prior to 1969." The Alabama bridges visited by HABS, including three in Colbert County, have all since been destroyed. Here is some information available from the surveys in the Library of Congress Digital Archives.

Big Bear Creek Covered Bridge was built in the mid-nineteenth century in Allsboro. It is shown here in 1936. *Courtesy of Historical American Buildings Survey/Library of Congress.*

BIG BEAR CREEK COVERED BRIDGE

Photographed by Alex Bush on April 29, 1936 Allsboro, Colbert County, spanning Big Bear Creek on County Road 7 Survey no.: HABS AL-361-A

Very little information can be found on this bridge. It was thought to have been built in the mid-1800s, but no date is recorded for its destruction. There are no covered bridges remaining in Colbert County. The Historic American Buildings Survey includes three exterior photos, as well as one taken from inside the bridge. It was in a severely dilapidated condition at the time and was missing part of its roof.

Cripple Deer Creek Covered Bridge was built in 1859 in Allsboro. It is shown here in the 1930s. *Courtesy of Historic American Buildings Survey/Library of Congress.*

CRIPPLE DEER CREEK COVERED BRIDGE

Photographed by Alex Bush on April 29, 1936 Cherokee, Colbert County, spanning Cripple Deer Creek River on County Road 1 Survey no.: HABS AL-361

Cripple Deer Creek Covered Bridge was built circa 1859. The survey includes five exterior photos and one interior photo. An exterior photo shows a detail of the bridge abutments made from stacked stone. The bridge was in dilapidated condition at the time.

Cripple Deer Creek Covered Bridge in Allsboro Colbert County, 1936. *Courtesy of Historic American Buildings Survey/Library of Congress.*

BUZZARD ROOST COVERED BRIDGE

Photographer unknown
Cherokee, Colbert County, Old Memphis Road
Survey no.: HABS AL-361-B

If its 1820 construction date is correct, this 94-foot span may have been Alabama's first covered bridge. However, most historians believe that it was actually built circa 1860. At the time of the HABS survey in 1935, the bridge was in bad shape and had a sign advertising a service station above its entrance. Sometime in the late 1960s or early 1970s, the bridge was given to the National Park Service to maintain as part of the Natchez Trace Parkway. It was destroyed by arson on July 15, 1972. The HABS survey includes only two exterior photos.

UNNAMED COVERED BRIDGE

Photographed by W.N. Manning on January 5, 1935
Eastaboga, Talladega County, spanning Choccolocco Creek on County Road 93
Survey no.: HABS AL-445

There is no name or date of construction listed with the bridge survey. Some online bridge listings have identified this as Lincoln Covered Bridge, built in 1903 and burned on July 15, 1963; however, the entrances do not appear to be of the same design. The survey includes five exterior photos and two interior photos.

A photo of Horton Mill Covered Bridge in Oneonta taken for the Historic American Engineering Record. *Courtesy of the Library of Congress.*

Left: Buzzard Roost Covered Bridge in Cherokee is shown in the 1930s. *Courtesy of Historic American Buildings Survey/Library of Congress.*

Below: Buzzard Roost Covered Bridge in Cherokee is shown in the 1930s. Its date of construction is unknown. Some historians say it was built as early as 1820. It burned in 1972. *Courtesy of Historic American Buildings Survey/Library of Congress.*

Two Alabama bridges were catalogued by HABS's sister program, the Historic American Engineering Survey, which was created in in 1969 by the NPS and the American Society of Civil Engineers. HAER's mission is to document historic mechanical and engineering artifacts, including bridges of all types. In Alabama, HAER surveyed Horton Mill Covered Bridge (HAER AL-203) and Swann Covered Bridge (HAER AL-201), both of which are still standing. The photos can be found in the Library of Congress Digital Archives.

THESE FAMILIES
BUILT THEIR OWN
COVERED BRIDGES

M ost covered bridges are owned by local governments or historical societies, but in some cases, people become so enamored of the picturesque bridges, they build one on their private property. In Alabama, several covered bridges are located on private property, and they come in a variety of styles and sizes. Following are the stories of three families who made their dreams of bridge ownership come true.

THE PRICES

In April 1963, Claude and Virginia Price of Elberta, Alabama, were headed to Niagara Falls for their honeymoon. It was the trendiest spot for newlywed couples at the time, and the Prices looked forward to witnessing the massive falls. They never arrived at their destination. Instead, Claude, now seventy-five years old and a retired barber, took a detour to look at some covered bridges. "We passed a covered bridge in Kentucky and I'd never seen one," Claude said. "I was amazed at the old carpentry they used."

The Prices stopped to see several other bridges while driving through Ohio. "Ohio had more than 600 covered bridges back then," he said. By that time, their arrival at Niagara Falls had been delayed, allowing time for a snowstorm to hit the area, closing the roads. "We never made it to Niagara Falls."

Price Covered Bridge was built by Claude Price in 2011 in Elberta. *Courtesy of Claude and Virginia Price.*

Claude's fascination with covered bridges had begun. "Every year for about 50 years, we went somewhere to see covered bridges," he said, adding that he has seen most of the ones in the United States. By 2010, Claude finally had a bridge of his own. Over the course of seven years and with the help of Horace York, Claude built a 72-foot-long, 27-foot-wide bridge over Mifflin Creek on his property on Selena Lane. It was constructed in a Queenpost Truss style.

In the years since their 1963 wedding, the Prices raised a family while also helping to foster numerous kids from St. Mary's Home, a refuge for abused and abandoned children. Although the couple no longer fosters children, they continue to help the organization as much as possible. In 2011, the Prices held a fundraiser for St. Mary's by hosting an event on his newly completed covered bridge. In exchange for donations to the home, visitors could walk through the bridge and be regaled with Claude's knowledge of covered bridges.

The Prices also rent the bridge, known simply as Price Covered Bridge, for weddings and family reunions. Claude said that in 2015, ninety-two Word War II veterans held a reunion at the bridge. The Prices donate any proceeds from rentals to St. Mary's, merging their two loves.

THE GILBERTS

In 1997, Wade Gilbert decided that his three-hundred-acre farm, Gilbert Ranch on Baker Hill Road in Limestone County, needed a covered bridge. It was built over a stream at a picturesque spot on the ranch where local Halloween events and Civil War reenactments were often held. The abutments of an old concrete bridge remained, giving Wade the idea to build his own covered bridge.

Wade, with the help of his sons—Grant, Joel and Thad—built the 40-foot-long, 12-foot-wide bridge in two weeks' time. The sides of the bridge, known as Gilbert Covered Bridge, were built of cherry and the floor of five-inch-thick elm boards.

THE OSBORNES

In 1985, Don and June Osborne were searching for property to buy in Limestone County, and they had one important consideration: a stream and enough land to build a covered bridge. "For years June and I have been interested in covered bridges," Don said. The couple had visited all of Alabama's historical bridges. "We decided we would like to have one built on a drive going into a house one day in the future."

They began searching. "When we started looking for property on which to build, we found the perfect lot because it had a stream running through the middle which would accommodate a bridge," Don said. "We had a difficult time convincing the owner of the property to sell it to us." The owner told the Osbornes the lot was "too low across the front to build on and one would have to build a bridge to get to the higher ground."

That's when Don told him "that was the very reason we wanted that lot." After purchasing the property, Don, a retired school principal, quickly bought three steel I-beams with the intention of building a bridge.

Years passed, and the I-beams still lay on the Osbornes' property off Dawson-Dupree Road. In 1998, Don learned of another local man who'd built a covered bridge on his property, Wade Gilbert. Don enlisted Wade's help, and soon they'd drawn plans for a 34-foot-long, 12-foot-wide span. The main portion of the bridge was built of cherry wood, while the flooring was beech. The bridge, named Creekwood Covered Bridge, was covered with a tin roof.

Creekwood Covered Bridge was built by Don and June Osborne in Limestone County. *Courtesy of Don Osborne.*

"Since then it has been our pleasure to have photographers bringing people for pictures with the bridge in the background," Don said. "We have given 2,700 4-by-6 colored pictures of the bridge with the history on the back and a final note to 'come drive though and bring your camera.'"

NATIONWIDE COVERED BRIDGE FESTIVALS

A labama is not the only state that celebrates the romance of its covered bridges with a festival. The list here is not meant to name every festival in the United States, but rather to highlight some festivals in several states. For more information, visit websites for each state's tourism department.

- Annual Covered Bridge and Arts Festival at Knoebels Amusement Resort, Elysburg, Pennsylvania, October
- Ashtabula County Covered Bridge Festival, Jefferson, Ohio, October
- Covered Bridge Arts and Music Festival, Oneonta, Alabama, October
- Covered Bridge Days, Brodhead, Wisconsin, August
- Elizabethton Covered Bridge Celebration, Elizabethton, Tennessee, June
- Euharlee Covered Bridge Fall Festival, Euharlee, Georgia, October
- Felton Covered Bridge Festival, Felton, California, Memorial Day weekend. This event is combined with the "Felton Remembers" parade, which honors those in the military.
- Madison County Covered Bridge Festival, Winterset, Iowa, October. This festival celebrates the bridges that were the subject of the book and film *The Bridges of Madison County*.
- Matthews Cumberland Covered Bridge Festival, Matthews, Indiana, September

- Oregon Covered Bridge Festival, Cottage Grove, Oregon, September
- Parke County Covered Bridge Festival, Rockville, Indiana, October. This is billed as the largest in the nation and features ten days of tours, activities, live music, arts and crafts and more.
- Roann Covered Bridge Festival, Roann, Indiana, weekend after Labor Day
- Union County Covered Bridge and Bluegrass Festival, North Lewisburg, Ohio, September
- Virginia Covered Bridge Festival, Woolwine, Virginia, June
- Washington & Greene Counties Covered Bridge Festival, Pennsylvania, September
- Westport Covered Bridge Festival, Westport, Indiana, June
- Zumbrota Annual Covered Bridge Music & Arts Festival, Zumbrota, Minnesota, June

SCENIC COVERED BRIDGE TOURS ACROSS THE COUNTRY

Here is just a sampling of the covered bridge tours you can take across the country.

- Bedford County, Pennsylvania. This driving tour passes fourteen historic covered bridges.
- Bucks County, Pennsylvania. This tour takes Bucks County visitors to twelve covered bridges. Bucks County once had more than fifty covered bridges; of the remaining twelve, ten are open to motor traffic.
- Bennington County, Vermont. Driving tour of five historic bridges.
- Covered Bridge Bicycle Tour, Albany, Oregon. This is one-day cycling event in the mid-Willamette Valley.
- Covered Bridges Electric Byway, Oregon. Twenty of Oregon's fifty covered bridges can be found in Lane County in the towns of Eugene, Springfield, Lowell and Cottage Grove. Most bridges are open to bicycles and pedestrians; a few are open to motor traffic.
- Historic Covered Bridges Driving Tour, Frederick County, Maryland. This tour takes visitors to three historic covered bridges listed in the National Register of Historic Places.
- Lancaster County, Pennsylvania. The county tourism department offers five driving tours, each including several historic covered bridges and ideas for things to do along the way.

- Lehigh Valley, Pennsylvania. Seven of the more than two hundred covered bridges in Pennsylvania are still standing. Five of these, all open to motor traffic, are showcased on this tour.
- Madison County, Iowa. Billed as the Covered Bridge Capital of Iowa, Madison County has the largest group of covered bridges in one area in the western half of the Mississippi Valley. The tour begins in Winterset.
- Marietta, Ohio. This driving tour passes nine surviving bridges of Washington County's original fifty covered bridges.
- Maysville, Kentucky. See eight of Kentucky's thirteen covered bridges on this five-county tour of northern Kentucky. The tour includes Fleming County's Goddard White Bridge. Built in the 1820s, it is the only surviving example of Ithiel Town's lattice truss design in the state.
- New England's Romantic Covered Bridges. The Alabama tourism department says, "New England has [covered bridges] by the score. Some are just for walking across a river; others are also for cycling. A surprising number are on roads, providing motorists with a brief 'back in time' experience."
- Parke County, Indiana. Parke County offers five driving tours showcasing thirty-one historic bridges, many built in the 1800s.
- Washington and Greene Counties Covered Bridge Tour, Pennsylvania. This driving tour takes visitors to thirty covered bridges.

PRESERVING THE NATION'S AND ALABAMA'S COVERED BRIDGES

Reference to a covered bridge often sparks an image of a quaint setting with a narrow, but inviting, timber tunnel crossing of a stream. For some, the image is of a more substantial structure crossing a raging river, withstanding the rigors of time and nature.
—*Federal Highway Administration's* Covered Bridge Manual

An interest in covered bridges and the desire to preserve them for future generations is probably one of the factors that led you to read this book. Perhaps this book has given insight into the importance of covered bridges to American culture and fueled a desire to help save bridges in your area.

The questions are: How do you help? Who is taking charge in preserving the relatively few bridges we have remaining in the United States? Here is a look at the core groups, organizations and legislation that are helping to preserve our covered bridge heritage in Alabama for future generations to cherish and enjoy.

THE NATIONAL SOCIETY FOR THE PRESERVATION OF COVERED BRIDGES

This nonprofit organization founded in 1950 has been instrumental in saving many of the country's historic covered bridges from demolition. According to the group's website at coveredbridgesociety.org, the mission of the organization is: "To preserve covered bridges. To gather and record knowledge of the history of covered bridges. To collect and preserve pictures, printed, and manuscript matter and other articles of historical interest concerning covered bridges. To do all things, alone or in cooperation with other persons or corporations, necessary or advisable to carry out any or all of the foregoing purposes and objects."

One of the group's largest contributions is publishing the *World Guide to Covered Bridges*. First printed in 1953 as a guide to historic covered bridges in Ohio, it now catalogues bridges around the world. The book uses a numbering system devised by the society and is used by historians and the Federal Highway Administration's *Covered Bridge Manual*.

FRIENDS OF THE COVERED BRIDGES OF BLOUNT COUNTY, ALABAMA

Friends of the Covered Bridges is a nonprofit organization formed in May 2011 by citizens concerned about the condition of Blount County covered bridges. Since then, the group has worked diligently to preserve its three historic bridges. Members worked with the Blount County Commission to obtain a federal grant issued by the National Historic Covered Bridge Preservation Program. The county added additional funding, and the group was able to restore the bridges and reopen them to motor traffic in 2012 and 2013.

Friends of the Covered Bridges arranges cleanup days to keep the bridges and surroundings beautiful by removing trash and overgrown shrubbery. In addition to restoring and maintaining the bridges, the group helped the Blount County Commission get surveillance cameras installed on the bridges to monitor the structures around the clock. This helps prevent vandalism and arson, one of the most common reasons for a bridge's demise.

While maintaining the bridges and keeping them safe from vandals are important goals, one of the group's biggest contributions is promoting tourism. People come from across the country to view the bridges of Blount County. The income is beneficial to the county and is one more reason to care for these historic structures. The Covered Bridge Arts & Music Fest, held each October in Oneonta, is the county's largest tourism draw.

Outside Alabama, a number of local groups like the Friends of the Covered Bridges of Blount County are working to protect bridges in

Members of the Blount County Memorial Museum gather for the opening of the Alabama Covered Bridge Trail. *Courtesy of Blount County Memorial Museum.*

their communities and states, and most are always looking for members. Preservation relies on the involvement of local citizens who care about their communities.

THE NATIONAL REGISTER OF HISTORIC PLACES

This listing is not only for historic homes. The National Register is the federal government's official list of districts, sites, structures and objects deemed to be of historical significance or worthy of preservation. The register was created under the National Historic Preservation Act of 1966. The legislation provided federal government oversight and control over historic preservation of historic structures for the first time. In its more than fifty-year history, the register has helped save or preserve more than eighty thousand individual historic sites.

Gargus Covered Bridge was built in 1966 in Gallant. *Photo by Wil Elrick.*

The National Register of Historic Places is administered by the National Park Service, which is overseen by the U.S. Department of the Interior. A nomination process is used for initial submission to the National Register, and during the nomination process, the property is evaluated in terms of the four criteria for inclusion in the National Register of Historic Places. The structure must have "significance in American history," as well as:

- be associated with events that have made a significant contribution to the broad patterns of our history; or
- be associated with the lives of significant persons in our past; or
- embody the distinctive characteristics of a type, period or method of construction, or that represent the work of a master, or that possess high artistic values or that represent a significant and distinguishable entity whose components may lack individual distinction; or
- have yielded, or may be likely to yield, information important in history or prehistory.

Many of the nation's oldest covered bridges have been added to the register, which is the most important way to preserve a historic covered bridge. If the bridge makes it through a grueling criteria process to be placed in the register, its care becomes eligible for certain tax incentives, as well as possible federal funding. If federal funding is used on a bridge project, the structure can't be demolished and must be maintained to a historical standard. Sadly, there have been numerous historic covered bridges that had to be removed from the register after they were destroyed by arson or natural disaster, including some in Alabama.

Alabama has several covered bridges placed in the National Register of Historic Places, and local governments and communities use the designation to keep these bridges safe. Alabama bridges on the national registry, as of April 2018, are:

- Clarkson-Legg Covered Bridge in Cullman County, built in 1904 and added to the registry on June 25, 1974.
- Coldwater Covered Bridge in Calhoun County, built circa 1845 and added to the registry on April 11, 1973.
- Easley Covered Bridge in Blount County, built circa 1927 and added to the registry on August 20, 1981.
- Horton Mill Covered Bridge in Blount County, built from 1934 to 1935 and added to the registry on December 29, 1970.
- Kymulga Mill and Covered Bridge in Talladega County, built circa 1860 (the mill) and circa 1864 (the covered bridge). They were added to the registry on October 29, 1976.
- Swan Covered Bridge in Blount County, built in 1933 and added to the registry on August 20, 1981.
- Nectar Covered Bridge in Blount County, built in 1934 and added to the registry on August 20, 1981. This bridge was destroyed by arson on June 13, 1993, and only the bridge supports remain. Even though it is currently still listed on the register, it will probably be removed.

Although Alabama has been losing historical bridges to neglect and arson, six covered bridges will soon be eligible to be added to the National Register. Structures must be at least fifty years old to qualify:

- Gargus Covered Bridge in Gallant (Etowah County), built in 1966. No application has been made for listing in the National Register of Historic Places. It is privately owned.
- Askew Covered Bridge in Auburn (Lee County), built in 1968. It is privately owned.
- Mountain Oaks Covered Bridge in Hoover (Jefferson County), built in 1970. It is located in a subdivision and is open to motor traffic.
- Fosters Covered Bridge in Dothan (Houston County), built in 1972. It is privately owned.
- Tannehill Valley Covered Bridge in McCalla (Jefferson County), built in 1972. It is located at the entrance of Tannehill Valley Estates Subdivision and is open to motor traffic.
- Cambron Covered Bridge in Huntsville (Madison County), built in 1974. It is located on a public nature trail on Green Mountain.
- Point Clear Covered Bridge in Fairhope (Baldwin County), built before 1975 at the historic Marriott Grand Hotel Point Clear Resort & Spa.

Top and middle: Coldwater Creek Covered Bridge in Oxford is shown during renovation. *Courtesy of Charlotte Kirkpatrick.*

Bottom: Twin Creeks Covered Bridge near Midway. *Courtesy of John Trent.*

THE FEDERAL HIGHWAY ADMINISTRATION

The Federal Highway Administration (FHWA) is an agency within the U.S. Department of Transportation that supports local and state governments in the design, construction and maintenance of the nation's highway system (under the Federal Aid Highway Program) and various federally and tribal owned lands (under the Federal Lands Highway Program). By offering financial and technical assistance to state and local governments, the Federal Highway Administration is responsible for ensuring that America's roads and highways are among the safest and most technologically sound in the world.

This agency is responsible for upholding the Transportation Equity Act for the 21st Century (TEA-21), which created the National Historic Covered Bridge Preservation Program. The program includes preservation of covered bridges that are listed, or are eligible for listing, in the National Register of Historic Places. It helps conduct research into better ways of restoring and protecting covered bridges. One project of the Covered Bridge Preservation Program was to create the FHWA-approved *Covered Bridge Manual*.

Consisting of 346 pages, the *Covered Bridge Manual* covers every conceivable topic that might help anyone trying to restore or maintain a covered bridge. It includes descriptions of bridge components, types of engineering and helpful resources. It also includes multiple case studies of existing bridge rehabilitation projects, as well as construction of new covered bridges.

LIST OF ALABAMA'S EXISTING COVERED BRIDGES

ALAMUCHEE COVERED BRIDGE, LIVINGSTON, SUMTER COUNTY

This bridge was built in the Town Truss style on the orders of Confederate army captain William Alexander Campbell Jones in 1861 and used to access Mississippi. Sometimes called Bellamy or Alamuchee-Bellamy, it originally spanned the Sucarnoochee River between Livingston and York. In 1924, it was moved about five miles from its original site to span Alamuchee Creek on the old Bellamy-Livingston Road. It was used for motor traffic until 1958, when it was abandoned. A historical marker at the site gives the year of construction as 1860. It reads, in part, "[In] 1860 Captain W.A.C. Jones of Livingston designed and built the bridge of hand-hewn yellow pine put together with large pegs, clear span 88 feet, overhead clearance 14 feet, and inside width 17 feet, across the Sucarnoochee River on old State Road South of Livingston." In 1971, the bridge was moved to the campus of the University of West Alabama, where it spans Duck Pond. Visitors can park in the Student Union parking lot and walk to the bridge.

AMANDA'S COVERED BRIDGE, SOUTHSIDE, ETOWAH COUNTY

This 12-foot-long covered bridge was built circa 1988 in the Stringer style and covered in 1998 in Southside. It is located on East Fowlers Ferry Road, half a mile from the junction of Alabama Highway 77. This bridge is on private property. Ask permission to visit.

ASKEW COVERED BRIDGE, AUBURN, LEE COUNTY

This 24-foot-long, Stringer-style bridge was built in 1968 by William K. Askew. It is located in the yard of a private home on North College Street in Auburn. Askew was a member of the Southern Covered Bridge Society. According to the book *Covered Bridges in the Southeastern United States: A Comprehensive Illustrated Catalog* by Warren H. White, a covered bridge where Askew's children played was demolished, and he wanted a bridge his children could continue to enjoy. He built Askew Bridge in his front yard and included diamond-shaped windows at adult height and at a child's height. It is only 4 feet wide. It is on private property at 543 Sanders Street. Ask permission to visit.

BLOUNT COUNTY MASTER GARDENERS MEMORIAL COVERED BRIDGE, ONEONTA, BLOUNT COUNTY

This bridge was built in Palisades Park in 2000 to honor the Master Gardeners of Blount County. It is 14 feet long and spans a ditch. The park also has a miniature covered bridge, as well as several pioneer buildings. The park is located at 1225 Palisades Parkway. For information, call 205-274-0017.

CAMBRON COVERED BRIDGE, HUNTSVILLE, MADISON COUNTY

This post and beam–style bridge was built by Madison County in 1974 along the Nature Trail atop Green Mountain. The 90-foot-long bridge crosses a cove of Sky Lake. It was named for Joe E. Cambron, who was the Madison County bridge foreman from 1958 until 1974. The Nature Trail is located at 5000 Nature Trail Road Southeast. For information, call 256-883-9501.

CLARKSON COVERED BRIDGE, CULLMAN, CULLMAN COUNTY

Also known as Clarkson-Legg Covered Bridge, the 250-foot-long span crosses Crooked Creek and is owned and maintained by the Cullman County Commission. It was built in 1904 in the Town Truss style. It is part of a park that includes a working gristmill and dogtrot cabin. Cullman County's website notes, "The bridge was torn in two, in 1921, by a huge storm. One piece was left intact, the other swept downstream and soon salvaged. One year later, the project to repair the bridge with the salvaged material was completed." A historical marker at the site reads, "Sometimes called Legg Bridge. This 270-foot bridge was constructed in 1904, destroyed by a flood in 1921 and rebuilt the following year. The only remaining covered bridge in Cullman County, it was restored by the Cullman County Commission in 1975 as an American Revolution Bicentennial Project. Named to Register of Historic Places, June 25, 1974."

The park is located at 1240 County Road 1043. For information, call 256-739-2916.

CLEARBRANCH UNITED METHODIST CHURCH COVERED BRIDGE, ARGO, JEFFERSON COUNTY

This 35-foot-long covered bridge was built in 2005 in Argo, near Trussville. The Stringer-style bridge is between Clearbranch United Methodist Church and its parking lot at 8051 Glenn Road.

COLDWATER COVERED BRIDGE, OXFORD, CALHOUN COUNTY

Sometimes called Hughes Mill Covered Bridge, this span is 63 feet long and was built no later than 1850—and probably earlier—making it the oldest covered bridge in Alabama. It was built in an unusual combination of methods: multiple Kingpost Truss with Town Lattice. It was listed in the National Register of Historic Places on April 11, 1973, under the name Coldwater Creek Covered Bridge. According to the Alabama Tourism Department, it was built by an unidentified slave. It originally spanned Coldwater Creek near Hughes Mill, about eight miles west of its current location. It was moved to Oxford Lake Park at 401 McCullars Lane and restored in 1990.

COOLEY COVERED BRIDGE, TRUSSVILLE, JEFFERSON COUNTY

This 27-foot-long, Stringer-style bridge was built in 2000 over Pinch Gut Creek in the Carrington Place Subdivision. It was built by Bob Smith and Roy Knighton. Ask permission to visit.

CREEKWOOD COVERED BRIDGE, ATHENS, LIMESTONE COUNTY

This 35-foot-long bridge in the Stringer style was built in 1998 over a brook on private property at 15370 Dawson Dupree Road. It was commissioned by Don Osborn and built by Wade Gilbert and his son, Joel. Ask permission to visit.

EASLEY COVERED BRIDGE, ONEONTA, BLOUNT COUNTY

This 83-foot-long bridge spans the Dub Branch of the Calvert Prong of the Little Warrior River. It was constructed circa 1927 in the Town Lattice-

Easley Covered Bridge was built in 1927 in Oneonta. *Photo by Wil Elrick.*

Truss style. Sometimes referred to as the Rosa Covered Bridge, the span is located on Easley Bridge Road off U.S. Highway 231, just south of the Rosa community.

Easley Bridge was listed in the Alabama Register of Landmarks and Heritage in 1978 and the National Register of Historic Places in 1981. Easley Bridge, one of three owned by Blount County, was in continuous use until it failed a 2009 inspection. It was restored and reopened to motor traffic in 2012. The National Historic Covered Bridge Preservation Program helped fund the restoration, as well as the renovation of Swann and Horton Mill Covered Bridges. In an effort to deter vandalism, the county installed cameras in all three bridges in 2015.

Foreman Forrest Tidwell led a crew that included his nephew, Zelma C. Tidwell, to build the bridge. Zelma would go on the build several other covered bridges in Alabama.

FOSTERS COVERED BRIDGE, DOTHAN, HOUSTON COUNTY

This 24-foot-long bridge in the Stringer style was built in 1972 over a pond. It is located behind 7583 Park Avenue. Ask permission to visit.

GABE'S COVERED BRIDGE, COTTONWOOD, HOUSTON COUNTY

This 49-foot-long span was built in 1994 on County Road 33 as an I-beam Stringer-style bridge. It was built to complement Gabe's Covered Bridge Restaurant, which is no longer in business. It is now on private property at 7990 County Road 33 South. Ask permission to visit.

GARGUS COVERED BRIDGE, GALLANT, ETOWAH COUNTY

Built in 1966, this 22-foot-long bridge spans Gargus Bass Lake. It was built in the Stringer style and is located on County Road 35/Gallant Road. From the Gallant Post Office, go north on Gallant Road for a quarter mile. The bridge is visible from the road. Ask permission to visit.

GIBSON COVERED BRIDGE, MUSCADINE, CLEBURNE COUNTY

This 15-foot-long bridge spans a pond outlet. The date of construction is unknown. It is located on County Road 49 near the junction of County Road 267. Ask permission to visit.

GILBERT COVERED BRIDGE,
ATHENS, LIMESTONE COUNTY

In 1997, Wade Gilbert built this bridge on his three-hundred-acre farm, Gilbert Ranch, located on Baker Hill Road in Limestone County. The 40-foot-long span was built atop the abutments of an old concrete bridge over a tributary of Big Creek. Wade constructed the bridge with the help of his sons Grant, Joel and Thad. Ask permission to visit.

GILLILAND COVERED BRIDGE,
GADSDEN, ETOWAH COUNTY

Sometimes known as Reece Bridge, or Gilliland-Reese (*sic*) Bridge, this 81-foot-long span was moved from its original location to Noccalula Falls Park in 1967. It was built in 1899 by Jesse Gilliland on his Reece City plantation. It spanned Big Wills Creek. When the bridge was donated to the City of Gadsden, it was rebuilt over a pond in the park where it is surrounded by other pioneer-style structures, including a gristmill. The centerpiece of the park is 90-foot waterfall that cascades into the gorge below. The park is located at 1500 Noccalula Road. For information, call 256-549-4663.

GOVERNORS PARK 1 AND 2,
HOLLY POND, CULLMAN COUNTY

Two green-and-white covered bridges span a pond in Governors Park located directly behind Holly Pond Town Hall at 10885 U.S. Highway 278 East. A 36-foot-long bridge crosses a small pond in the center of the park. A 13-foot-long bridge spans a small inlet from the pond. Both were built in 1998 in the Stringer style. For information, call 256-796-2124.

One of two green-and-white covered bridges spans a pond in Governors Park located directly behind Holly Pond Town Hall. *Photo by Wil Elrick.*

HOOTERS COVERED BRIDGE, DOTHAN, HOUSTON COUNTY

This bridge has no official name; its name was derived from its location in the parking lot of an abandoned Hooters restaurant at 3385 Ross Clark Circle. The picturesque green-and-white bridge spans a drainage ditch. The 36-foot-long bridge was built in 1999 in the Stringer style.

HORACE KING MEMORIAL BRIDGE, VALLEY, CHAMBERS COUNTY

This bridge was built in 2003 in the style of former slave and bridge builder Horace King. It was built on an old railroad trestle located on the

A view of Horace King Memorial Bridge in Valley. *Courtesy of Shannon Kazek.*

Chattahoochee Valley Railroad Trail on U.S. Highway 29. The 31-foot-long bridge spans Moores Creek. Call 334-756-5228 for information.

HORTON MILL COVERED BRIDGE, ONEONTA, BLOUNT COUNTY

Rising 70 feet above the water, this is the highest covered bridge in the nation. The 203-foot-long bridge spans the Calvert Prong of the Little Warrior River. Built in 1934 by Clyde Tidwell in the Town Truss style, this bridge was restored in 2013. Surveillance cameras were installed in 2015

A view of Horton Mill Covered Bridge in Oneonta. *Photo by Wil Elrick.*

to prevent vandalism. The Horton Mill Bridge was listed in the National Register of Historic Places on December 29, 1970. It is open to motor traffic, but because it has only one lane, drivers must show courtesy to other drivers and observe a five-mile-per-hour speed limit. The bridge is located just off Alabama Highway 75, five miles north of Oneonta. For information, call 205-274-2153.

HUGH KING COVERED BRIDGE, SPRINGVILLE, ST. CLAIR COUNTY

This 44-foot-long bridge was built in 1986 over Little Canoe Creek in Springville. The Stringer-style bridge is located on private property at 1680 Alabama Highway 174. Ask permission to visit.

INDIAN COVERED BRIDGE,
MENTONE, DEKALB COUNTY

This 14-foot-long, Stringer-style bridge crosses a gorge on Alabama Highway 117, 1.4 miles from the junction with County Road 89.

IVALEE COVERED BRIDGE,
ATTALLA, ETOWAH COUNTY

This I-beam Stringer-style bridge is 51 feet long and spans Brown Creek on a private drive off U.S. Highway 278. It was built in 2005. Ask permission to visit.

A view up the driveway through Ivalee Covered Bridge in Attalla. *Photo by Kelly Kazek.*

JEFF BARTON COVERED BRIDGE,
SUMITON, WALKER COUNTY

Built in 1994 over Mathis Creek on the Barton property, this bridge is 45 feet long. Ask permission at the Barton residence on Mathis Creek Farm Road. Travel west on U.S. Highway 78, turn south on Mathis Creek Farm Road and continue for two-tenths of a mile, then turn southeast on Barton Place (the driveweay to the residence).

JIMMY LAY COVERED BRIDGE,
EMPIRE, WALKER COUNTY

This 21-foot-long, King-style covered bridge was built in 1984 over a tributary of Sloan Creek. It is located off Empire Road, two-tenths of a mile off Main Street. It is visible from the road. Ask permission to visit.

KYMULGA COVERED BRIDGE,
CHILDERSBURG, TALLADEGA COUNTY

Kymulga Grist Mill and Covered Bridge are owned by the City of Childersburg and open to the public as a historic park. The mill was built in 1864 and is one of the few gristmills that survived the Civil War, when most were burned. The Kymulga Bridge, built in 1861, is one of only two nineteenth-century bridges that remain in their original locations. Waldo Covered Bridge is the other. The 105-foot-long bridge was built in the Howe Truss style over Talladega Creek. It is maintained by the City of Childersburg. The bridge located at 7346 Grist Mill Road. Call 256-378-7436 for information.

Mason Covered Bridge is a 32-foot-long bridge built in 1997 in Cleveland. *Photo by Wil Elrick.*

MASON COVERED BRIDGE, CLEVELAND, BLOUNT COUNTY

Built in 1997, this 32-foot-long bridge spans a brook off Alabama Highway 160 at the junction of Alabama Highway 79. It was built in the Stringer style.

MOTHER'S COVERED BRIDGE, MILLBROOK, ELMORE COUNTY

This 40-foot-long bridge built in 1999 spans Coosada Creek. It is located on private property near 5500 Willow Tree Drive. It was constructed in the Stringer style by Jim Lee, Gee Eason and James Calhoun. It was toppled by a tornado in 2000 and rebuilt by the original builders. The entrance is barred with a cast-iron gate marked "Private Property." Ask permission to visit.

MOUNTAIN OAKS ESTATES COVERED BRIDGE, HOOVER, JEFFERSON COUNTY

This bridge spans Huckleberry Creek on Mountain Oaks Road in the Mountain Oaks Estates Subdivision. The 26-foot-long bridge was built in 1970 with a post-supported roof.

OLD DOWNING MILL COVERED BRIDGE, CHOCCOLOCCO, CALHOUN COUNTY

This bridge was built over Egoniaga Creek in the I-beam, Stringer style in 1993. The 95-foot-long bridge is on private property at 1350 Old Mill Road. Ask permission to visit.

OLD UNION CROSSING COVERED BRIDGE, MENTONE, DEKALB COUNTY

The original portion of Old Union Crossing Bridge is said to have been built in 1863, although this has not been confirmed. Only the center part of the bridge, located on a dirt road that dead-ends on the Old Military Trail, is covered. That 42-foot-long portion was moved to the site in 1972 and set atop an existing bridge, which is 90 feet long. It spans the West Fork of the Little River. The privately owned bridge is located on Lookout Mountain not far from the ski slopes at Cloudmont Ski & Golf Resort off DeKalb County Road 614 in Mentone.

OVERLAND ROAD BRIDGE, BRIERFIELD, BIBB COUNTY

This bridge was built in 1994 over Furnace Brook in Brierfield Ironworks Historical State Park. The 47-foot-long bridge of Stringer construction features the face of a playful troll above its entrance. This bridge was originally located in Montevallo in Shelby County and is sometimes referred to as Montevallo Bridge. Brierfield Park's centerpiece is the ruin of a 36-foot-tall brick furnace built by Bibb County Iron Company in 1862 to produce iron for farm implements. In 1863, under pressure from the Confederate government,

A troll decorates Overland Road Covered Bridge in Brierfield Ironworks Park. *Photo by Kelly Kazek.*

the owners sold the ironworks to the Confederacy for $600,000. The park is located at 240 Furnace Parkway, Brierfield, Alabama, 35035. The park includes other historic attractions and allows camping. For information, call 205-665-1856, or visit www.brierfieldironworks.com.

PERRY LAKES PARK COVERED BRIDGE, MARION, PERRY COUNTY

This 80-foot-long bridge spans a stream in Perry Lakes Park. Its unusual peaked-roof design was created by architecture students in Auburn University's Rural Studio program. It is covered with a tin roof. The park features several other structures built by the Rural Studio, including a100-foot-tall birding tower, restrooms and a pavilion. Bald eagles have been known to nest in the 125-acre park, which is made up of four oxbow lakes formed more than 150 years ago when the Cahaba River changed its course. The park is located off Alabama Highway 175, five miles east of Marion near the State Fish Hatchery. For information, call 334-247-2101.

POINT CLEAR BRIDGE, POINT CLEAR, BALDWIN COUNTY

This 51-foot-long bridge was built at the historic Marriott Grand Hotel Resort and Spa sometime before 1975. The Stringer-style bridge spans Point Clear Creek. The hotel is located at 1 Grand Boulevard in Point Clear. Call 251-928-9201 for information.

POOLE'S COVERED BRIDGE,
TROY, PIKE COUNTY

In 1998, Wyndel Eiland built this 60-foot-long bridge to be part of the Pioneer Museum of Alabama. It was constructed in the Town Lattice Truss style over a pond. It was named for Grover Poole, a local logger who for many years hauled out timber with his horse. The park also features replica pioneer buildings and antique farm equipment. The collection includes cabins, a gristmill, a church, a depot, a steam train and more. It is located at 248 U.S. Highway 231 North. For information, call 334-566-3597.

PRICE COVERED BRIDGE,
NEAR ELBERTA, BALDWIN COUNTY

This 72-foot-long bridge was built by Claude Price in 2011 over Miflin Creek on his property. The Queen-style bridge with steel construction is on Price's private driveway off Main Street, one-tenth of a mile from the junction with U.S. Highway 98. Ask permission to visit.

PUMPKIN HOLLOW COVERED BRIDGE,
STERRETT, SHELBY COUNTY

Built in 1992 over Bear Creek, this 68-foot-long bridge is located behind 18274 Rocky Hollow Lane, an unmarked road, in the gated community of New Pumpkin Hollow Lake. The Town Lattice on the bridge is decorative and not part of the structure's support. Ask permission to visit.

RICHARD MARTIN TRAIL COVERED BRIDGES
NO. 1 AND 2,
ELKMONT, LIMESTONE COUNTY

Two existing bridges on the Richard Martin Trial off Piney Chapel Road near Elkmont were converted in 2017 to covered bridges. The trail follows an abandoned railway, passing the site of the Civil War Battle of Sulphur Creek Trestle. One bridge is located 1.2 miles north of the Piney Chapel

Road trailhead, while the other is 1.4 miles north of the Upper Fort Hampton Road trailhead. For information, call Limestone County Parks and Recreation Department at 256-216-3425.

RIKARD'S MILL COVERED BRIDGE, BEATRICE, MONROE COUNTY

This 81-foot-long bridge is one of the most unusual in Alabama because it's actually a gift shop for Rikard's Mill park, which includes a circa 1840s gristmill, a syrup mill, a blacksmith shop and a replica pioneer cabin. The bridge was constructed atop an old steel bridge in 1994 with a shop built on top. The mill was constructed by Jacob Rikard, a local blacksmith. It was operated by the Rikard family until the 1960s. The family eventually donated the mill to the Monroe County Heritage Museums. The mill was placed in the Alabama Register of Landmarks and Heritage in 1998. The park was closed to the public for a portion of 2017, so be sure to call before going. It is located at 4116 Alabama Highway 265 North. Call 251-789-2781 for updated information.

ROMINE COVERED BRIDGE, ATHENS, LIMESTONE COUNTY

This 35-foot-long bridge was constructed in 2004 over Beauchamp Branch. It is located on private property behind 16495 Zehner Road. Ask permission to visit.

SALEM-SHOTWELL COVERED BRIDGE, OPELIKA, LEE COUNTY

This 43-foot-long Town Lattice Truss–style bridge was rebuilt in 2007 after the original 1900 bridge was destroyed by a falling tree. It was initially built by Otto Puls over Wacoochee Creek on Shotwell Road/County Road 252 near the Salem community. It was also known as the Pea Ridge Covered Bridge. The original bridge, which was 75 feet long, was made using longleaf heart pine, white oak pegs and cedar shakes. After it was destroyed, it was rebuilt from salvaged materials in Opelika Municipal Park, where it is now

Romine Covered Bridge is located on Zehner Road in Limestone County. Built in 2004, the 35-foot-long bridge spans Beauchamp Branch. It is private property, so permission required. *Courtesy of Don Osborne.*

Salem-Shotwell Covered Bridge was built in Lee County in 1900. It was rebuilt in Opelika in 2007 after it was heavily damaged by a falling tree. *Courtesy of Shannon Kazek.*

maintained by the City of Opelika. The bridge was listed in the Alabama Register of Landmarks and Heritage on January 25, 1977. The bridge is located on Park Road. For information, call 334-705-5560.

SANDAGGER COVERED BRIDGE, THEODORE, MOBILE COUNTY

This 11-foot-long covered bridge was built in 2007 as a portion of a much longer bridge. It was built by Fred Sandagger in the Stringer style at the residence of 11320 Bellingrath Road. The bridge, including uncovered portions, has seven spans and measures 79 feet in total. Ask permission to visit.

SAUNDERS FAMILY COVERED BRIDGE, STERRETT, SHELBY COUNTY

This 51-foot-long bridge over a spillway of Lake Laurlee is located on the property of the now-defunct Twin Pines Resort. Built in 1988, it is sometimes referred to as Twin Pines Covered Bridge. It is located on private property at 1200 Twin Pines Road. Ask permission to visit.

SWAN CREEK WALKING TRAIL COVERED BRIDGE, ATHENS, LIMESTONE COUNTY

Built in 2004 to accompany a city walking trail, this bridge is 40 feet long and built in the Stringer style. It spans Swan Creek. There are two entrances to the trail from the highway; the closest to the bridge begins at Athens Sportsplex at 1403 U.S. Highway 31.

SWANN COVERED BRIDGE, CLEVELAND, BLOUNT COUNTY

Also known as the Swann-Joy Covered Bridge, this 324-foot-long span is the second longest in Alabama, after Twin Creeks Covered Bridge in Bullock County. Built in 1933, Swann Bridge spans the Locust Fork of the Warrior River. It was built on the Swann Farm to provide access to the Joy

Swan Creek Covered Bridge on the walking trail in Athens. *Photo by Wil Elrick.*

A photo of Swann Covered Bridge in Cleveland, Alabama, taken for the Historic American Engineering Record. *Courtesy of the Library of Congress.*

community. It is one of three covered bridges owned and maintained by the Blount County Commission. It was restored in 2012 and reopened to motor traffic. Surveillance cameras were installed in 2015 to prevent vandalism. Swann Covered Bridge was listed in the National Register of Historic Places in 1981. It is located on Swann Bridge Road, 1.5 miles from the junction with U.S. Highway 231.

TANNEHILL VALLEY ESTATES COVERED BRIDGE, MCCALLA, JEFFERSON COUNTY

Built in 1972, this bridge marks the entrance to the Tannehill Valley Estates Subdivision. It is 42 feet long and built in the Stringer style. It crosses Tannehill Mill Stream and is located on Eastern Valley Road.

TWIN CREEKS COVERED BRIDGE, NEAR MIDWAY, BULLOCK COUNTY

This bridge, the longest in the state, is 334 feet long and spans two creeks at the Wehle Land Conservation Center. The Stringer-style bridge, located about three miles into the park, is built over a concrete slab and rests on forty-four wooden supports. It was built by Tom Hall and is open to motor traffic. A plaque on the bridge reads, "Twin Creek Covered Bridge. The bridge was built in the summer of the year 2000. Its 334 feet is believed to be the fifth longest covered bridge in the United States and the longest one in Alabama. The primary purpose of covered bridges in the south is to protect the quality of the wood from weather. The wood for this bridge was harvested from Baldwin County in southeast Alabama. Martin Hicks cut and readied the wood for construction. Tom Hall from Tuscaloosa County constructed the bridge. The Cypress Trees cut had completed their life cycle so cutting was not environmentally harmful. The design was taken from the bridges built in the late 18th century." The center is located at 4890 Pleasant Hill Road in Midway. For information, call 334-529-3003.

Tannehill Valley Estates Covered Bridge was built in 1972 in McCalla. It marks the entrance to the Tannehill Valley Estates subdivision. *Courtesy of Charlotte Kirkpatrick.*

Twin Creeks Covered Bridge near Midway is the longest in the state at 334 feet. It spans two creeks at the Wehle Land Conservation Center. *Courtesy of John Trent.*

VALLEY CREEK PARK COVERED BRIDGE, SELMA, DALLAS COUNTY

This Stringer-style bridge was built in 1989 by the City of Selma. The 82-foot-long bridge spans Valley Creek. The park is adjacent to Bloch Park, which includes a field named for Major League Baseball player Terry Leach, a Selma native. It is located at 108 West Dallas Avenue. For information, call 334-874-2140.

WALDO COVERED BRIDGE, TALLADEGA, TALLADEGA COUNTY

This bridge, built circa 1858 and the second oldest in Alabama, is in imminent danger of collapse. It is the only one of the state's historical bridges that is not being maintained. The approaches to this 116-foot-long bridge are missing, so there is no access.

Also known as Riddle Mill Covered Bridge, this span of combined Howe Truss/Queenpost Truss construction was built on a Socopatoy Indian Trail near Riddle's Mill, a gristmill that was later made into the town hall for the tiny burg of Waldo and then a restaurant. The mill is currently empty. The bridge is also near Riddle's Hole, a gold mine that was in operation for more than a century, from 1840 through World War II. Waldo Bridge was used by Wilson's Raiders in 1865. The bridge is eligible for inclusion in the National Register of Historic Places.

A PERSONAL PLEA

Covered bridges are strong, capable structures meant to withstand the test of time, and with the proper care, many have. But too many of our historic covered bridges have been lost to neglect, vandalism and arson. While I didn't intend to inject personal feelings into this book, I decided that my experience at the ruins of a historic bridge would illustrate what we lose when a covered bridge is destroyed.

In preparing this book, I visited Blount County Memorial Museum in Oneonta to scour its archives and get information on local bridges. While talking with historian Laura Roberson, I asked for directions to the ruins of the Nectar Covered Bridge, once the longest in the state and seventh longest in the nation at 385 feet, so I could take photos. She told me that it had purposefully been set on fire in 1993. When she heard the news, she rushed to the scene to find the structure completely destroyed. After twenty-five years, the weight of the loss was still obvious in her demeanor.

I followed her directions to the small community of Nectar. The bridge ruins had long been on my list of places to visit because I recalled hearing the catastrophic news of Nectar Bridge's demise when I was teenager growing up in Marshall County, Alabama. Driving over the concrete bridge that replaced the beautiful covered bridge a short distance away, I could easily see the remaining abutments. Getting down to the riverbed with my camera proved more of a challenge, but I managed to shimmy down the steep bank of the Locust Fork of the Black Warrior River.

These stone abutments are what remain of Nectar Covered Bridge, which was the longest in the state at 385 feet before it burned in 1993. *Photo by Wil Elrick.*

When I got to the riverbed, I settled my feet in the sand, took stock of the gently rolling brown waters and got ready to take what promised to be some amazing photographs. But I found that I couldn't, at least not right away. As I looked at the two supports rising from water, I was transported to the bridge's heyday, imagining when it was an integral part of this tiny community. As I pictured cars or carriages driving across the span, I could feel the same sense of loss I had seen in Laura Roberson's eyes. This was the moment that I truly understood the fascination with covered bridges.

The thought of this structure, a feat of engineering, being deliberately destroyed infuriated me. I knew I was not alone in this feeling, and that mighty Nectar Covered Bridge was not the only bridge to be a victim of arson. It is one of the leading causes of destruction of covered bridges. For me, the best way to help save these great structures was to write a book about them, so I managed to take several photos of the Nectar ruins to accompany this book. I hope that others will participate in the revival of their endangered bridges, such as Waldo Covered Bridge in Talladega County, Alabama, which has already lost its approaches and is in danger of collapse.

—WIL ELRICK

BIBLIOGRAPHY

PREFACE

The Bridges of Madison County. Director Clint Eastwood, Warner Bros., 1995.
Encyclopedia Britannica. "Covered Bridge, Engineering." Brittanica.com.
National Society for the Preservation of Covered Bridges.
New Hampshire Tour Guide. "Why Are Covered Bridges Covered?" Compiled
    and edited by Richard G. Marshall. New Hampshire Department of
    Transportation, 1994.
Round Barns and Covered Bridges. "US Covered Bridge Lists." DaleJTravis.com.
The Straight Dope. "Why Are Covered Bridges Covered?" StraightDope.com.
U.S. Department of Transportation, Federal Highway Administration.
    Covered Bridge Manual. April 2005, publication FHWA-04-098.
Waller, Robert James. *The Bridges of Madison County*. N.p.: Warner Books, 1992.
Wikipedia. "Covered Bridge." https://en.wikipedia.org/wiki/Covered_bridge.

WHY ARE BRIDGES COVERED?

AL.com. "Preserving the Romance of Alabama's Covered Bridges." July 18,
    2013.
Spartanburg (SC) Herald. "Alabama Landmarks Pass." July 13, 1936.
U.S. Department of Transportation, Federal Highway Administration.
    Covered Bridge Manual.

ALABAMA'S COVERED BRIDGE
TRAIL AND FALL FESTIVALS

Alabama Department of Tourism.
Blount County Chamber of Commerce.
Blount County Memorial Museum.
Covered Bridge Arts and Music Fest.

HORACE KING: FROM SLAVE
TO RENOWNED BRIDGE BUILDER

Alabama Department of Archives and History.
Columbus State University.
Encyclopedia of Alabama.
ExploreGeorgia.org.
Lupold, John S., and Thomas L. French. *Bridging Deep South Rivers: The Life and Legend of Horace King*. Athens: University of Georgia Press, 2004.
New Georgia Encyclopedia.
Trust for Public Land.

LEGENDS ABOUT COVERED BRIDGES

American Motorcyclist Magazine (April 1988).
The Guardian. "Longest Covered Bridge in World in N.B. Made for Kissing and Luring Tourists." October 2011.
Laughlin, Robert W.M., and Melissa C. Jurgensen. *Kentucky's Covered Bridges*. Charleston, SC: Arcadia Publishing, 2007.
VisitSleepyHollow.com.

THE HANGING AT ALAMUCHEE BRIDGE

CraigRenfroe.com.
Kazek, Kelly, and Wil Elrick. *Alabama Scoundrels: Outlaws, Pirates, Bandits and Bushwhackers*. Charleston, SC: The History Press, 2014.

MILLER BRIDGE WAS ONCE LONGEST IN STATE

BhamWiki.com.
Horseshoe Bend National Military Park.
National Park Service, U.S. Department of the Interior. https://www.doi.
gov/hurricanesandy/nps.

WAS THIS BRIDGE A CIVIL WAR CROSSING?

White, Warren H. *Covered Bridges in the Southeastern United States.* Jefferson, NC:
McFarland, 2012.

RURAL STUDIO'S UNIQUE COVERED BRIDGE AND HOUSE

Ruralstudio.org.
RuralSWAlabama.org.

ALABAMA'S LOST COVERED BRIDGES

Alabama Department Archives and History Digital Collections. http://
digital.archives.alabama.gov.
Alabama Mosaic. http://alabamamosaic.org.
National Park Service, U.S. Department of the Interior. https://www.doi.
gov/hurricanesandy/nps.
Round Barns and Covered Bridges. "US Covered Bridge Lists: Alabama."
DaleJTravis.com.

GRISTMILLS AND COVERED BRIDGES MAKE PICTURESQUE PAIRS

Cullman County Parks and Recreation.
KymulgaGristMill.com.
Noccalula Falls State Park.
Pioneer Museum of Alabama.
Rikard's Mill Historical Park.

TYPES OF COVERED BRIDGES

Bridges and Tunnels of Allegheny County & Pittsburgh, PA. "Bridge Basics."
pghbridges.com.
The Federalist. "Covered Bridges and the Birth of American Engineering."
Review. Society for History in the Federal Government newsletter
(Summer 2016).
Maryland Covered Bridges. "Covered Bridge Trusses." mdcoveredbridges.
com.
National Society for the Preservation of Covered Bridges. "Truss Types."
U.S. Department of Transportation, Federal Highway Administration.
Covered Bridge Manual.

RECORD-BREAKING BRIDGES AND OTHER TRIVIA

Horseshoe Bend National Military Park.
National Park Service, U.S. Department of the Interior. https://www.doi.
gov/hurricanesandy/nps.
National Society for the Preservation of Covered Bridges.
Wehle Land Conservation Center.
World Guide to Covered Bridges. A bridge numbering system by John Diehl,
this was developed in 1953 but is now updated regularly by the National
Society of Covered Bridges. http://coveredbridgesociety.org/downloads/
wg-update.pdf.

TALES OF HAUNTINGS AT ALABAMA'S COVERED BRIDGES

Alabama Haunted Houses. "Oakachoy Covered Bridge." AlabamaHaunted Houses.com.

Ghost Village. "Shotwell Covered Bridge." GhostVillage.com.

Haunted Places. "Oakachoy Covered Bridge." HauntedPlaces.org.

The Moonlit Road. "Haunted Bridge of Lookout Mountain, Alabama." TheMoonlitRoad.com.

Wikipedia. "Oakachoy Covered Bridge." https://en.wikipedia.org/wiki/Oakachoy_Covered_Bridge.

BRIDGES DOCUMENTED BY THE HISTORIC AMERICAN BUILDINGS SURVEY

Historic American Buildings Survey. http://www.loc.gov/pictures/collection/hh.

Library of Congress Digital Archives. https://www.loc.gov/collections.

THESE FAMILIES BUILT THEIR OWN COVERED BRIDGES

Gilbert, Wade. Article in the *(Athens, AL) News Courier*, 1998.

Hoffman, Roy. "Retired Barber Who Built Covered Bridge Uses It for St. Mary's Home Fundraiser." AL.com, March 24, 2011.

Osborne, Don. Interview by Kelly Kazek, January 11, 2018.

Price, Claude, and Virginia Price. Interview by Kelly Kazek, February 5, 2018.

NATIONWIDE COVERED BRIDGE FESTIVALS

Annual Covered Bridge Celebration Facebook page.

Ashtabula County Covered Bridge Festival Facebook page.

Blount County Memorial Museum in Oneonta, Alabama.

Covered Bridge Days-Brodhead, Wisconsin Facebook page.

CoveredBridgeFestival.com.

CoveredBridges.com.

Covered-Bridges.org.

ElizabethtonChamber.com.

Eugene, Cascades & Oregon Coast Visitor Information. "Cottage Grove to Host Oregon Covered Bridge Festival." September 18, 2014. https://www.eugenecascadescoast.org.

Euharlee Covered Bridge Fall Festival Facebook page.

Euharlee.com/events.

FeltonFestival.com.

Felton Remembers Parade and Covered Bridge Festival Facebook page.

ItourColumniaMontour.com.

Knoebels.com.

Madison County Covered Bridge Festival Facebook page.

MadisonCounty.com.

MatthewsCoveredBridgeFestival.com.

MtnsofMusic.com.

ParkeCountyGuide.com.

RoannCoveredBridgeFestival.com.

TravelWisconsin.com.

Virginia Covered Bridge Festival Facebook page.

Washington & Greene Counties Covered Bridge Festival Facebook page.

Westport Indiana Covered Bridge Festival Facebook page.

WestportIndiana.org.

Zumbrota Annual Covered Bridge Music & Arts Festival Facebook page.

SCENIC COVERED BRIDGE TOURS ACROSS THE COUNTRY

CityofMaysville.com.

DiscoverNewEngland.com/things-do.

GreeneCountyTourism.org.

MadisonCounty.com/the-covered-bridges.

MariettaCounty.org/plan-a-trip.

MVBC.com (Mid-Valley Bicycle Club).

TravelOregon.com.
VisitBedforCounty.com/CoveredBridges.
VisitFrederick.org.
VisitVermont.com.

PRESERVING THE NATION'S AND ALABAMA'S COVERED BRIDGES

Alabama Department of Archives and History Digital Collections. http:// digital.archives.alabama.gov.

Blount County Memorial Museum. https://blountmuseum.org/index.html.

National Society for the Preservation of Covered Bridges. CoveredBridgesSociety.org.

———. "World Guide to Covered Bridges." http://www. coveredbridgesociety.org/wg.htm.

Round Barns and Covered Bridges. "US Covered Bridge Lists: Alabama." DaleJTravis.com.

LIST OF ALABAMA'S EXISTING COVERED BRIDGES

Alabama Department of Archives and History Digital Collections. http:// digital.archives.alabama.gov.

Bridges to the Past: Alabama's Historical Bridges. Documentary film, director Max Shores, 1999.

Round Barns and Covered Bridges. "US Covered Bridge Lists: Alabama." DaleJTravis.com.

AFTERWORD

Rhudy, Amy. Interview by Wil Elrick, Blount County Memorial Museum, February 4, 2018.

INDEX

ABOUT THE AUTHORS

KELLY KAZEK is a journalist, humor columnist and blogger. She is the author of nine other books, including two humor collections. She has won more than 180 state and national press awards and twice served as president of the Alabama Associated Press Media Editors. She lives in Huntsville, Alabama, with her husband, Wil Elrick. They travel Alabama's backroads together, seeking out quirky history, and have coauthored two books.

WIL ELRICK hails from Guntersville in the northeastern part of Alabama. He is a writer and "weirdologist" who loves telling stories, whether as a tour guide to historic Huntsville or with friends around a campfire. He can often be found "off the beaten path" researching historical, weird or unusual tales. He lives in Huntsville, Alabama, with his wife, Kelly Kazek, with whom he has coauthored two books.